Chemical Process Calculations

LECTURE NOTES

K ASOKAN

Universities Press

CRC Press
Taylor & Francis Group
Boca Raton London New York

CRC is an imprint of the Taylor & Francis Group, an informa business

CHEMICAL PROCESS CALCULATIONS

Universities Press (India) Private Limited

Registered Office
3-6-747/1/A and 3-6-754/A Himayatnagar,
Hyderabad 500 029, Telangana, India
email: info@universitiespress.com

Distributed by
Orient Longman Private Limited

Registered Office
3-6-752 Himayatnagar, Hyderabad 500 029, Telangana, India

Other Offices
Bengaluru, Bhopal, Chennai, Guwahati, Hyderabad, Jaipur, Kolkata, Lucknow, Mumbai, New Delhi, Noida, Patna, Vijayawada

Reprinted 2012, 2017, 2018

First published in India by
Universities Press (India) Private Limited 2007

ISBN: 978-81-7371-594-5

501436

Typeset by
Krishtel eMaging Solutions Pvt. Ltd., Chennai 600 017

Printed at
The Print Park, Chennai

Published by
Universities Press (India) Private Limited
3-6-747/1/A and 3-6-754/A Himayatnagar,
Hyderabad 500 029, Telangana, India
email: info@universitiespress.com

Dedication

This book is dedicated to my dear wife and children. Their comforting presence and willingness to sacrifice their time with me enabled me to concentrate fully on my work.

Preface

Chemical Engineers guide the passage of a product from the laboratory to the market place, from idea or prototype to a functioning article or process, from theory to reality. They design and operate large-scale chemical process equipment and units safely and efficiently, and in an environmentally responsible manner. They produce a diverse range of materials from fuels and fertilisers to processed foods, life-saving pharmaceuticals and filtered clean water. All this requires a thorough understanding of the physical and chemical aspects of materials as well as the energy balance of the processes involved. A chemical engineer can get this knowledge by taking a course on chemical process calculations.

The subject matter of this book is treated in eight chapters divided on conceptual basis. In each chapter, the fundamental principles are explained in simple language, followed by illustrations for clarity. At the end of each chapter, problems have been given for the learner to solve. In all there are 173 worked examples and 154 exercise problems.

The book is intended as a textbook for students taking a diploma in chemical technology, undergraduate chemical engineering and biotechnology, and postgraduate course in applied chemistry.

Karaikudi
July 2007

K Asokan

Acknowledgements

I would like to express my deep sense of gratitude to the following people who have all made this book possible: Professor A K Shukla, Director, Central Electrochemical Research Institute, Karaikudi for his encouragement; my colleague, Dr K Subramanian, for his mental, moral and physical help during the course of preparation of the book, and Mr S Sakthivel, for his neat and patient typing of the material.

Karaikudi
July 2007

K Asokan

Contents

1 Dimensions, Units and Conversions

Nearly every engineering problem encountered involves dimensions: the *length* of a house, the *mass* of a vessel, the *velocity* of a projectile, the *force* of the air resistance on a motor car, and so on. These *dimensions* can be expressed using specific *units*. For example, length can be expressed in feet, mass as kilograms, velocity as miles per hour, and force as newtons.

The objective of this chapter is to explain the use of dimensions and units in engineering calculations, and to introduce the standard systems of units that are used.

FUNDAMENTAL QUANTITIES AND DERIVED QUANTITIES

Dimensions are used to describe physical quantities. Physical quantities like length, mass, time, electric current, temperature and luminous intensity are called *fundamental quantities*. All other physical quantities like area, volume, velocity, acceleration, force and others, which can be expressed in terms of the fundamental quantities are called *derived quantities*. For example, area is length times length, volume is area times length, velocity is length divided by time, acceleration is velocity divided by time, force is mass times acceleration and so on.

FUNDAMENTAL UNITS AND DERIVED UNITS

The process of determining the number of times a quantity is bigger than a standard quantity is known as measurement. Thus measurement means comparison with a certain standard of the same kind. The accepted standard that is used for comparison of a physical quantity is known as a *unit*. Units in which fundamental quantities are measured are called *fundamental units*. The units of derived physical quantities are expressed in terms of the fundamental units and are called *derived units*.

ENGLISH SYSTEM OF UNITS

The common system of units used in the United States is the English system. In the English system of units, the unit of length is the foot, the unit of mass is the pound mass, and the unit of time is the second It is also known as the **FPS** system. Units are arbitrary, based on a standard that is developed and agreed upon within a system. Most English system units were inherited from the Romans. For example, the unit of length comes from the average length of a man's foot, and the mile was the average distance a Roman soldier could travel in 1000 paces.

The English system is confusing for several reasons. First, the unit of force is also called the pound, and is defined as one pound mass times the standard acceleration of gravity. It uses non-decimal conversion constants such as 12 inches per foot, 5 280 feet per mile, 43 560 square feet per acre, and 231 cubic inches per gallon.

SI SYSTEM OF UNITS

A more rational, decimal-based system of units was developed in France during the 1700s and was adopted by France in 1795. This system was called the **Metric System**, from the Greek word meaning 'to measure', and has become the standard system in science in much of the world. The standard form is now known as the International System or the **SI system**: The basic unit of length is the metre, the basic unit of mass is the kilogram, and the basic unit of time is the second. The main advantage of this system is that the conversion constants to all other units within the system are decimals. Today there is an international standard for the metre, the kilogram, the second, the ampere of electricity and the candela of light.

SI Base Units

The International System has seven base units from which nearly all the other units are derived.

Table 1.1 SI Base Units

Physical quantity	Name of unit	Symbol
Length	metre	m
Mass	kilogram	kg
Time	second	s
Electric current	ampere	A
Temperature	kelvin	K
Amount of substance	mole	mol
Luminous intensity	candela	Cd

SI Derived Units

The International System has numerous derived units of which many have special names (Table 1.2).

SI Prefixes

In general, numbers should be expressed using appropriate multipliers such that they fall between 0.1 and 1000; 1 250 m is best written as 1.25 km (Table 1.3).

Table 1.2 SI Derived Units

Physical quantity	Name of SI unit	Symbol of SI unit	In SI base units
Force	newton	N	$= kg\ m\ s^{-2}$
Pressure	pascal	Pa	$= kg\ m^{-1}\ s^{-2}$
Energy	joule	J	$= kg\ m^2\ s^{-2}$
Power	watt	W	$= kg\ m^2\ s^{-3}$
Charge	coulomb	C	$= A\ s$
Frequency	hertz	Hz	$= s^{-1}$

Table 1.3 SI Prefixes

Multiple	Prefix	Symbol	Multiple	Prefix	Symbol
10^{-1}	deci	d	10	deca	da
10^{-2}	centi	c	10^2	hecto	h
10^{-3}	milli	m	10^3	kilo	k
10^{-6}	micro	μ	10^6	mega	M
10^{-9}	nano	n	10^9	giga	G
10^{-12}	pico	p	10^{12}	tera	T
10^{-15}	femto	f	10^{15}	peta	P

Rules and Conventions in the Use of SI Units

1. The units named after a scientist should not be written with a capital initial letter.
   Example: 10 newtons not 10 Newtons; 5 watts not 5 Watts
2. The symbol for a unit named after a scientist should be written with a capital initial letter.
   Example: 20 N not 20 n
3. The symbol for all other units should not be written with a capital letter.
   Example: m for metre and not M
4. Only the singular form of the unit is to be used.
   Example: 5 kg and not 5 kgs
5. No fullstop or any other punctuation mark should be used within or at the end of the unit.
   Example: 100 m and not 100 m. 50 kg m^{-3} and not 50 kg m^{-3}.
6. When temperature is indicated as kelvin, the degree sign is to be omitted.
   Example: 373 K and not 373°K
7. Space is to be left between the numeral and symbol and between symbols of compound units such as velocity, acceleration, and others.
   Example: 30 s and not 30s; 22 m s^{-1} and not 22 ms^{-1}
8. Use of solidus is to be avoided; when used, not more than one solidus should be employed.
   Example: m s^{-2} or m/s^2 and not m/s/s

9. Units must be written in full or using agreed symbols, but no other abbreviation can be used.
   Example: Amp is not a valid abbreviation for ampere
10. When numerical values fall outside the range of 0.1 to 1000, the numerals should be separated into groups of three with a space replacing the traditional comma.
    Example: 5 000 and not 5,000 or 5000; 0.030 456 and not 0.030,456 or 0.030456

CONVERSIONS

In a widely expanding world, familiarity with the various systems of units and an ability to convert from one unit to another are essential. In scientific work, length, mass and time are the three fundamental quantities of interest. In engineering, force is an additional fundamental quantity to be taken into consideration. The unit of force in the FPS system is the poundal. It is the force which will give a mass of one pound, an acceleration of 1 ft s^{-2}. The unit of force in the SI units is newton. It is the force, which will give a mass of one kilogram, an acceleration of 1 m s^{-2}.

It is good to be familiar with the concepts of SI units. Whenever, a problem is encountered with FPS units, they can be converted to SI units they can. From this point of view, the conversion factors acquire importance.

Conversion to SI Units

Basic units

Length: 1 foot = 0.304 8 m

Mass: 1 pound = 0.453 6 kg

Force: 1 poundal = 0.138 N

$$\left(\text{Temperature in °F} - 32) \times \frac{100}{180}\right) + 273 = \text{temperature in kelvin}$$

Freezing point of water, 32°F = 273 K

Normal boiling point of water, 212°F = 373 K

1 K = 1.8°F

Derived units

Area: 1 ft^2 = 0.093 m^2

Volume: 1 ft^3 = 0.028 3 m^3

1 US gallon = 3.786 L

Pressure 1 inch Hg = 3.39 kN m^{-2}

$$1 \text{ psi} = 6.895 \text{ kN m}^{-2}$$

$$\begin{aligned} 1 \text{ atm} &= 101\,325 \text{ N m}^{-2} \\ &= 101.325 \text{ kN m}^{-2} \\ &= 1.013\,25 \text{ bar} \\ &= 1.033 \text{ kgf cm}^{-2} \\ &= 760 \text{ Torr} \end{aligned}$$

Energy: $1 \text{ Btu} = 1\,055 \text{ J}$

Power : $1 \text{ HP} = 748 \text{ W}$

ILLUSTRATIONS

1A The temperature inside an oven is 500°F. Report the temperature in SI units.

$$\text{Temperature in kelvin} = (500 - 32) \times \frac{5}{9} + 273$$

$$= 533 \text{ K}$$

1B The specific heat of a metal is 2 Btu $\text{lb}^{-1}\,°\text{F}^{-1}$. Find the value in SI units.

$$\text{The specific heat of the metal} = 2 \text{ Btu} \times \frac{1\,055 \text{ J}}{\text{Btu}} \times \frac{1}{\text{lb}} \times \frac{\text{lb}}{0.453\,6 \text{ kg}} \times \frac{1}{°\text{F}} \times \frac{1.8°\text{F}}{1 \text{ K}}$$

$$= 8\,373 \text{ J kg}^{-1} \text{ K}^{-1}$$

1C 2.5 US gallons of gasoline weigh 16.69 lb. Find the weight of 2 litres of gasoline in kilograms.

$$\text{Weight of gasoline per litre} = 16.69 \text{ lb} \times \frac{0.453\,6 \text{ kg}}{\text{lb}} \times \frac{1}{2.5 \text{ gallon}} \times \frac{1 \text{ gallon}}{3.78 \text{ L}}$$

$$= 0.799\,9 \text{ kg L}^{-1} \approx 0.8 \text{ kg L}^{-1}$$

The weight of 2 litres of gasoline is 1.6 kg.

1D Convert a heat transfer coefficient value of 1000 Btu ft^{-2} h^{-1} °F^{-1} into SI units.

$$1000\ \text{Btu ft}^{-2}\ \text{h}^{-1}\ {}^\circ\text{F}^{-1} = 1000\ \text{Btu} \times \frac{1\,055\ \text{J}}{\text{Btu}} \times \frac{1}{\text{ft}^2} \times \frac{\text{ft}^2}{0.093\ \text{m}^2}$$

$$\times \frac{1}{\text{h}} \times \frac{\text{h}}{3\,600\ \text{s}} \times \frac{1}{{}^\circ\text{F}} \times \frac{1.8{}^\circ\text{F}}{\text{K}}$$

$$= 5\,672.04\ \text{J m}^{-2}\ \text{K}^{-1}$$

1E The latent heat of vapourisation of saturated steam at 250 psi is 1 201.1 Btu lb^{-1}. Convert the pressure and enthalpy values into SI units.

$$\text{Pressure} = 250\ \text{psi} \times \frac{6.895\ \text{kN m}^{-2}}{\text{psi}}$$

$$= 1\,723.75\ \text{kN m}^{-2}$$

$$\text{Enthalpy} = 1\,201.1\ \text{Btu} \times \frac{1\,055\ \text{J}}{\text{Btu}} \times \frac{1}{\text{lb}} \times \frac{\text{lb}}{0.453\,6\ \text{kg}}$$

$$= 2\,793.56\ \text{kJ kg}^{-1}$$

Problems for Practice

1.1 Make the following conversions

(i) 142 ft^2 h^{-1} to SI units
(ii) 50 psi to kN m^{-2}
(iii) 100 Btu to kJ
(iv) 1000 W to HP
(v) 1000 lb ft^{-3} to SI units
(vi) 100°F to kelvin

1.2 Air flows through a duct at the rate of 200 lb ft^{-2} h^{-1}. Get the air flow rate in kg m^{-2} s^{-1}.

1.3 The heat transfer coefficient in the case of a drying operation is given by the dimensional equation.

$$h_y = 0.01\frac{G^{0.8}}{D_e^{0.2}}$$

where h_y = heat transfer coefficient in Btu $ft^{-2}\ h^{-1}\ °F^{-1}$

G = mass velocity in lb $ft^{-2}\ h^{-1}$

D_e = equivalent diameter of the air flow channel in feet

Convert the equation to SI units.

1.4 The mass transfer coefficient for a gas absorption operation is given by $kg_a = 0.028\, G_y^{0.7} G_x^{0.25}$ where kg_a is the gas phase mass transfer coefficient in mol $ft^{-3}\ h^{-1}\ atm^{-1}$. G_y is the gas rate and G_x is the liquid rate in lb $ft^{-2}\ h^{-1}$ respectively. Transform the expression into SI units.

1.5 Fouling factors for ordinary industrial liquids fall in the range of 300 to 1000 Btu $ft^{-2}\ h^{-1}\ °F^{-1}$. Give this range in SI units.

2 Basic Concepts

The basic chemical principles required for a clear understanding of material and energy balance will be discussed in this chapter. Matter exists in three different forms: solid, liquid and gas. The simplest mode of expressing the quantity of matter is mass. Solids and liquids can be quantified by weighing them in a balance. The volume of liquids can be measured and if their densities are known, their weight can be determined. A gas occupies the entire volume available and quantification requires a knowledge of temperature and pressure. A common unit used to quantify all the three forms of matter is the *mole*.

Amount of Substance The *mole* is the SI unit of amount of substance. It is defined as the amount of substance that contains as many elementary entities as there are atoms in 0.012 kg of carbon. In other words, one mole contains N elementary entities or basic species, where N is the Avogadro number (6.023×10^{23}).

Volume The *cubic metre* is the SI unit of volume, while *litre* is the common unit of volume. One litre is the volume occupied by one kilogram of air-free water at 277 K, the temperature at which its density is maximum.

$$1 \text{ litre (L)} = 1 \text{ dm}^3 = 1000 \text{ cm}^3$$

$$1 \text{ m}^3 = 1 \text{ kL} = 1000 \text{ L}$$

Force The *newton* is the SI unit of force. One newton is the force which when applied to a body of mass 1 kg gives it an acceleration of 1 m s^{-2}.

Pressure Pressure is defined as force acting on unit area. The SI unit of pressure is N m^{-2} or *pascal* (Pa). The *atmosphere* is the common unit of pressure. Pressure is usually measured with the help of a gauge which denotes the difference in the pressure of the vessel and the local atmospheric pressure. This is known as P_g or gauge pressure. Gauge pressure does not reflect the actual pressure of the vessel. To get the actual or absolute pressure (P), atmospheric pressure has to be added to the gauge pressure.

$$\text{Absolute pressure} = \text{Gauge pressure} + \text{atmospheric pressure}$$

$$P = P_g + \text{atmospheric pressure}$$

1 atmosphere gauge (1 atm g) denotes 2 atmospheres absolute, as the atmospheric pressure has to be added to the gauge pressure, as mentioned above. The atmospheric pressure in different units is given below.

$$\begin{aligned}
1 \text{ atmosphere} &= 101\,325 \text{ N m}^{-2} \\
&= 101\,325 \text{ Pa} \\
&= 101.325 \text{ bar} \\
&= 760 \text{ mm Hg} \\
&= 760 \text{ Torr} \\
&= 76 \text{ cm Hg} \\
&= 10.33 \text{ m } H_2O \\
&= 1.033 \text{ kgf cm}^{-2}
\end{aligned}$$

If no letter follows the pressure unit, absolute pressure has to be presumed. Often, pressure is expressed as pressure head.

$$\text{Pressure head} = \frac{\text{absolute pressure}}{\text{density of the barometric liquid}}$$

The common barometric liquids are mercury and water. One atmosphere pressure is equal to 76 cm mercury column or 1033 cm (10.33 m) water column. To convert the height of a mercury column to that of water column, the height of the mercury column has to be multiplied by the specific gravity of mercury, which is 13.6.

Vacuum refers to sub-atmospheric pressure. It is denoted by Torr or Pa or m bar.

Work and power Work is the product of force acting on a body and the distance travelled by the body under its influence. *Joule* is the SI unit of work. When a force of one newton acts on a body and the body moves through a distance of 1 m, the work done is said to be one joule or 1 J.

Power is the rate of doing work. The *watt* is the SI unit of power. $1 \text{ W} = 1 \text{ J s}^{-1}$

Temperature The unit of temperature in the SI system of units is *kelvin* (K). It is 1/273 of the triple point of water, which is 273 K.

$$\text{The normal freezing point of water} = 32°\text{F} = 0°\text{C} = 273 \text{ K}$$

$$\text{The normal boiling point of water} = 212°\text{F} = 100°\text{C} = 373 \text{ K}$$

A difference of 100 on the Celsius scale = 100 on the Kelvin scale = 180 on the Farenheit scale.

∴ One degree on the Kelvin scale = one degree on the Celsius scale = 1.8 degrees on the Farenheit scale.

$$\text{Temperature in farenheit} = \left(\text{Temperature in kelvin} \times \frac{9}{5}\right) + 32$$

$$\text{Temperature in kelvin} = (\text{Temperature in celsius} + 273)$$

Molecular weight The molecular weight is the weight of one mole of a substance. It is also equal to the weight of all the constituent elements of a compound present in one mole.

Atomic weight The atomic weight is the weight of one mole of an element.

Empirical formula The empirical formula shows the ratio of atoms of each element present in a compound.

Example: CH for benzene.

Molecular formula The molecular formula shows the actual number of atoms of each element present in a compound. It is n times the empirical formula, where n is a whole number.

Example: C_6H_6 for benzene.

Equivalent weight The equivalent weight is the number of parts by weight of a substance which will either react with or release in a reaction 1.008 parts by weight of hydrogen, or 8 parts by weight of oxygen, or 35.46 parts by weight of chlorine or one equivalent of any other substance.

Average molecular weight When a system comprises of more than one substance, then the average molecular weight has to be taken into consideration. It is the algebraic sum of the products of mole fraction and molecular weight of each of the components.

$$m = \Sigma m_i x_i$$

where m is the average molecular weight, m_i is the molecular weight of the ith component; and x_i is the mole fraction of the ith component.

The mole fraction of a component is the ratio of the number of moles of that component (n_i) to the total number of moles (n) in the system.

$$x_i = \frac{n_i}{n}$$

If a system comprises of 1 mol of nitrogen and 3 mol of hydrogen, then the mole fraction of nitrogen is 0.25 and that of hydrogen is 0.75. It may also be noted that the sum of the mole fractions of the components of a system is one: $\Sigma x_i = 1$

Boyle's Law Boyle's law states that at constant temperature, the pressure (P) of a given mass of gas is inversely proportional to its volume (V). At constant temperature, $P \alpha \frac{1}{V}$.

Charles' Law Charles' law states that at constant pressure, the volume (V) of a given mass of gas is directly proportional to its absolute temperature (T). At constant pressure, $V \alpha T$.

Combining Boyle's Law and Charles' Law, for 1 mol of a gas

$$PV \alpha\ T$$
$$PV = RT$$

For a system comprising of n moles of gas, the above equation becomes

$$PV = nRT$$
$$R = \frac{PV}{nT}$$

where R is the universal gas constant. Its value varies with the units in which P and V are expressed.

$$\begin{aligned} R &= 8.314\ \text{N m mol}^{-1}\ \text{K}^{-1} \\ &= 8.314\ \text{Pa m}^3\ \text{mol}^{-1}\ \text{K}^{-1} \\ &= 8.314\ \text{J mol}^{-1}\ \text{K}^{-1} \\ &= 0.082\,05\ \text{lit atm mol}^{-1}\ \text{K}^{-1} \\ &= 0.082\,05\ \text{m}^3\ \text{atm kmol}^{-1}\ \text{K}^{-1} \\ &= 1.987\ \text{cal mol}^{-1}\ \text{K}^{-1} \\ &= 1.987\ \text{kcal kmol}^{-1}\ \text{K}^{-1} \end{aligned}$$

Density The density of a gas is the mass per unit volume of the gas.

$$\text{Density} = \frac{\text{mass}}{\text{volume}}$$

If we consider one mole of the gas, the mass of the gas is its molecular weight (M) and its volume is the molar volume (V), then,

$$\text{Density} = \frac{\text{molecular weight}}{\text{molar volume}} = \frac{M}{V}$$

Making use of the universal ideal gas equation, V can be replaced by $\frac{RT}{P}$. Substituting this value of V in the above equation, we have

$$\text{Density} = \frac{PM}{RT}$$

Using this equation, the density of a gas can be calculated at any given temperature and pressure. The commonly used unit of density is g L^{-1} or kg m^{-3}.

The mass flow rate of the gas can be computed, if the volumetric flow rate of the gas and its density are known.

$$\text{Mass flow rate } (q_m) = \text{volumetric flow rate} \times \text{density}$$

Specific volume The specific volume of a gas is the volume per unit mass of the gas. In otherwords, specific volume is the reciprocal of density.

Specific gravity The specific gravity of a gas is the ratio of its density to that of air, at the same temperature and pressure.

$$\text{Specific gravity} = \frac{\text{density of gas}}{\text{density of air}}$$

If we consider one mole of the gas, density of the gas is the ratio of its molecular weight to molar volume and density of air is the ratio of its average molecular weight to molar volume. As we know that the molar volume of all gases are equal at the same temperature and pressure,

$$\text{Specific gravity} = \frac{\text{molecular weight of the gas}}{\text{average molecular weight of air}}$$

It may be noted that since air is a mixture of nitrogen and oxygen gases, its molecular weight should be denoted as average molecular weight.

ILLUSTRATIONS

2A Find the density of nitrogen at 303 K and 2 atm gauge pressure.

Solution

2 atm gauge pressure is equal to 3 atm absolute pressure.

Density of the gas can be found by the formula, density $= \dfrac{PM}{RT}$

Molecular weight of nitrogen is 28. When pressure is substituted in atmospheres and volume in litres, the value of R to be used is 0.082 05 lit atm mol^{-1} K^{-1}.

$$\text{Density of nitrogen at 303 K and 3 atm pressure} = \frac{3 \times 28}{0.082\,05 \times 303}$$

$$= 3.38 \text{ g L}^{-1} \text{ or } 3.38 \text{ kg m}^{-3}$$

2B A gas contains 81.8% carbon and 18.2% hydrogen by weight. If 369 ml of this gas at 293 K and 748 Torr weigh 0.66 g, what is the molecular formula of the gas?

Solution

The compound contains two elements carbon and hydrogen. Their weight per cent is given. If their atomic ratio is calculated, the empirical formula, which is the ratio of atoms of each element present in a compound, can be obtained.

Weight ratio of carbon to hydrogen is $81.8 : 18.2$

If the weight of the element is divided by the atomic weight, the atomic ratio is obtained.

$$\text{Atomic ratio} = \frac{81.8}{12} : \frac{18.2}{1} = 6.82 : 18.2 = 1 : 2.667$$

or $3:8$ ($\because$ the ratio should be a whole number)

Empirical formula of the gas = C_3H_8

If volume (V), pressure (P) and temperature (T) of a gas are known, the number of moles of gas (n) present can be found by the formula, $n = \dfrac{PV}{RT}$.

$$\text{Number of moles present in 369 ml of gas} = \frac{748}{760} \times \frac{0.369}{0.08205 \times 293}$$

$$= 0.015 \text{ mol}$$

When pressure is substituted in atmospheres and volume in litres, the value of R to be used is 0.082 05 lit atm mol^{-1} K^{-1}.

Weight of 0.015 mol = 0.66 g

Weight of 1 mol = 44 g

That is, the molecular weight of the compound is 44.
Formula weight as per empirical formula is also 44. Hence the empirical formula and molecular formula are one and the same, that is, C_3H_8.

2C Find the volume of 88 kg of carbon dioxide at a pressure of 2 atm gauge pressure and 373 K.

Solution

Molecular weight of carbon dioxide = 44

Amount of carbon dioxide in 88 kg of gas (n) $= \dfrac{88}{44} = 2$ kmol

Pressure on the absolute scale (P) $= 2 + 1 = 3$ atm

Volume of the gas can be found by the formula $V = \dfrac{nRT}{P}$

When pressure is substituted in atmospheres and volume in litres, the value of R to be used is 0.082 05 lit atm mol^{-1} K^{-1}.

$$\text{Volume of the gas} = \frac{2 \times 0.082\,05 \times 373}{3}$$

$$= 20.4\ m^3$$

2D Find the pressure exerted by 132 kg of carbon dioxide if it is confined to a 40 m^3 container at 573 K.

Solution

Molecular weight of carbon dioxide $= 44$

Amount of carbon dioxide in 132 kg of gas (n) $= \frac{132}{44} = 3$

Volume of the container (V) $= 40\ m^3$

Temperature (T) $= 573$ K

Pressure of the gas can be found by the formula $P = \frac{nRT}{V}$

$$= \frac{3 \times 0.082\,05 \times 573}{40}$$

$$= 3.53 \text{ atm} = 2.53 \text{ atm g}$$

$$= 357\,282\ N\ m^{-2}$$

2E Find the temperature reached by 176 kg of carbon dioxide when compressed to 10 bar pressure and confined to a 30 m^3 container.

Solution

Molecular weight of carbon dioxide $= 44$

Amount of carbon dioxide in 176 kg of gas (n) $= \frac{176}{44} = 4$

Pressure (P) $= \frac{10}{1.013}$ atm

Volume of the gas (V) $= 30\ m^3$

Temperature of the gas can be found by the formula $T = \frac{PV}{nR}$

$$= \frac{10 \times 30}{1.013 \times 4 \times 0.082\,05}$$

$$= 902.35\ \text{K}$$

2F Find the weight of LPG of composition 60% propane and 40% butane when confined to a 3 m^3 cylinder at 8 atm gauge pressure at room temp. Assume that the weight of the cylinder is 11.13 kg.

Solution

8 atm gauge pressure is equal to 9 atm absolute pressure.

Volume of the gas (V) $= 3\ m^3$

Temperature of the gas (T) $= 303$ K

Number of moles of the gas can be found by the formula $n = \frac{PV}{RT}$

$$= \frac{9 \times 3}{0.082\,05 \times 303} = 1.086\ \text{kmol}$$

As volume is in cubic metres, n will be in kilomoles.

LPG is a mixture of gases. Its molecular weight has to be found by using the formula for average molecular weight.

$$m = \Sigma m_i x_i$$

where m is the average molecular weight, m_i is the molecular weight of the ith component, and x_i is the mole fraction of the ith component.

For gaseous mixtures, the valid assumption for the composition is volume or molar basis. Mole per cent given in the problem has to be transformed to mole fraction by dividing it by 100 (as mole per cent is out of 100 and mole fraction is out of one).

Average molecular weight of LPG $= (0.6 \times 44) + (0.4 \times 58)$

$= 49.6$

Weight of LPG = number of moles × average molecular weight

$= 1.086 \times 49.6$

$= 53.87$ kg

Weight of the empty cylinder $= 11.13$ kg

Weight of the cylinder with LPG $= 53.87 + 11.13$

$= 65$ kg

$$\text{Number of cylinders that can be loaded on a 10 tonne truck} = \frac{10\,000}{65} = 153.85$$

$$\approx 154 \text{ cylinders}$$

Dalton's law The total pressure of a mixture of gases is the sum of the partial pressures of the component gases.

$$P = \Sigma P_i$$

Pure component pressure or partial pressure (P_i) of a gas is the pressure that it will exert if it is the only gas present in the whole of the container.

Amagat's law The total volume of a mixture of gases is the sum of the partial volumes of the component gases.

$$V = \Sigma V_i$$

Pure component volume or partial volume (V_i) is the volume that a component will occupy if present alone at that temperature and pressure.

Raoult's law The partial pressure of a component in the vapour space above a solution of a liquid solute in a liquid solvent is directly proportional to the mole fraction of that component in the solution.

$$P_A \ \alpha \ x_A$$

$$P_A = P_A^* x_A$$

The proportionality constant P_A^* is the vapour pressure of the liquid solute, P_A is the partial pressure of the liquid solute in the vapour space above the liquid solution, and x_A is the mole fraction of the solute in the solution. Similarly,

$$P_B \ \alpha \ x_B$$

$$P_B = P_B^* x_B$$

The proportionality constant P_B^* is the vapour pressure of the liquid solvent, P_B is the partial pressure of the liquid solvent in the vapour space above the liquid solution, and x_B is the mole fraction of the solvent in the solution.

For a two component system,

$$x_A + x_B = 1$$

$$x_B = (1 - x_A)$$

$$P_B = P_B^*(1 - x_A)$$

Rauolt's law holds good for a liquid–liquid mixture or when both solute and solvent are in the liquid state. Solute denotes the component present in minor amount and solvent denotes the component present in major amount.

Henry's law The partial pressure of the gaseous solute in the vapour space above a solution of the gaseous solute in a liquid solvent is directly proportional to the mole fraction of the solute in the solution.

$$P_A \;\; \alpha \;\; x_A$$

$$P_A = Hx_i$$

where, H is the proportionality constant known as Henry's law constant. P_A is the partial pressure of the gaseous solute in the vapour space above the solution, and x_A is the mole fraction of the solute in the solution.

Henry's law holds good for a liquid–gas mixture or when the solute is a gas and the solvent is a liquid.

To prove that partial pressure (P_A) = mole fraction (N_A) × total pressure (P)

Consider a three component system of gases, A, B and C. Let P be the pressure, V the volume and T the temperature.

According to the ideal gas equation

$$P_A V = n_A RT$$

$$P_A = \frac{n_A}{V} RT \tag{1}$$

$$P_B = \frac{n_B}{V} RT \tag{2}$$

$$P_c = \frac{nc}{V} RT \tag{3}$$

Adding equations (1), (2) and (3)

$$P_A + P_B + P_C = (n_A + n_B + n_C)\frac{RT}{V} \tag{4}$$

By Dalton's Law,

$$P = P_A + P_B + P_C$$

$$P = (n_A + n_B + n_C)\frac{RT}{V} \tag{5}$$

Dividing equation (1) by equation (5),

$$\frac{P_A}{P} = \frac{nA}{nA + nB + nC}$$

$$= N_A$$

$$P_A = N_A P$$

To prove that partial volume (V_A) = mole fraction (N_A) × total volume (V)

Consider a three component system of gases A, B and C. Let the total volume be V and pressure and temperature be P and T.

According to the ideal gas equation

$$PV_A = n_A RT$$

$$V_A = n_A \frac{RT}{P} \quad (1)$$

$$V_B = n_B \frac{RT}{P} \quad (2)$$

$$V_C = n_C \frac{RT}{P} \quad (3)$$

Adding equations (1), (2) and (3)

$$V_A + V_B + V_C = (n_A + n_B + n_C) \frac{RT}{P} \quad (4)$$

By Amagat's Law,

$$V = V_A + V_B + V_C$$

$$V = (n_A + n_B + n_C) \frac{RT}{P} \quad (5)$$

Dividing equation (1) by equation (5),

$$\frac{V_A}{V} = \frac{n_A}{n_A + n_B + n_C}$$

$$= N_A$$

$$V_A = N_A V$$

ILLUSTRATIONS

Note:

1. Problems should be solved by choosing a basis. Proper choice of basis makes the solving of the problem much easier.
2. The basis may be a period of time or a given mass of material.

2G When nitrogen tetroxide is heated to 673 K, 50% of it dissociates. If 30 kg of nitrogen tetroxide is heated to 673 K and 740 mm Hg, calculate the volume occupied by the resulting gases. ($N_2O_4 \rightleftharpoons 2\,NO_2$).

Solution

Basis: 30 kg of nitrogen tetroxide

Molecular weight of nitrogen tetroxide $= 92$

Amount of nitrogen tetroxide initially present $= \frac{30}{92} = 0.326$ kmol

On heating, 50% of nitrogen tetroxide dissociates. Each molecule of nitrogen tetroxide gives 2 molecules of nitrogen dioxide on dissociation.

Amount of undissociated nitrogen tetroxide $= 0.326 \times 0.5 = 0.163$ kmol

Amount of nitrogen dioxide formed $= 0.163 \times 2 = 0.326$ kmol

Amount of gas present at equilibrium $= 0.489$ kmol

Volume of the gases can be found using the ideal gas equation $V = \frac{nRT}{P}$

When pressure is reported in atmospheres, the value of R to be used is 0.082 05 m^3 atm $kmol^{-1}$ K^{-1}.

Volume of the gases at 673 K and 740 mm Hg $= \frac{0.489 \times 0.08205 \times 673 \times 760}{740}$

$= 27.76\ m^3$

2H In the gaseous state nitrogen tetroxide partially dissociates according to the equation $N_2O_4 \rightleftharpoons 2\,NO_2$. If 24 g of a mixture of nitrogen tetroxide and nitrogen dioxide occupy a volume of 15 100 cm^3 at 363 K and 730 mm Hg pressure, calculate the percentage dissociation of nitrogen tetroxide.

Solution

Basis: 24 g of nitrogen tetroxide

Molecular weight of nitrogen tetroxide $= 92$

Amount of nitrogen tetroxide initially present $= \frac{24}{92} = 0.261$ mol

When heated nitrogen tetroxide dissociates to nitrogen dioxide and the gases occupy a volume of 15.1 L at 363 K and 730 mm Hg pressure. Number of moles of the gases in this volume can be found using the ideal gas equation, $n = \frac{PV}{RT}$. When pressure is reported in atmospheres, the value of R to be used is 0.082 05 L atm mol^{-1} K^{-1}.

$$\text{Amount of nitrogen tetroxide after dissociation} = \frac{730 \times 15.1}{760 \times 0.082\,05 \times 363}$$
$$= 0.487 \text{ mol}$$

Let x moles of nitrogen tetroxide dissociate and form $2x$ moles of nitrogen dioxide.

Undissociated nitrogen tetroxide $= 0.261 - x$

Nitrogen dioxide formed $= 2x$

Amount of gases after dissociation $= (0.261 - x) + 2x$

$= 0.261 + x$

$\therefore \quad 0.261 + x = 0.487$

$x = 0.226$ mol

$$\text{Percentage of dissociation} = \frac{0.226}{0.261} \times 100 = 86.7\%$$

2I The flow rate of air from a blower is to be determined. The blower delivers air at 323 K and 750 mm Hg pressure. 10 kg of ammonia are added every minute to the air. A sample of the gaseous mixture taken downstream indicates 67.1% nitrogen, 17.9% oxygen and 15% ammonia. Compute the flow rate of air delivered by the blower in cubic metres per minute.

Solution

Basis: One minute operation

Weight of ammonia added to air $= 10$ kg

Molecular weight of ammonia $= 17$

$$\text{Amount of ammonia added to air} = \frac{10}{17} = 0.588 \text{ kmol}$$

Ammonia forms 15% of the gaseous mixture after mixing. Air accounts for the balance of 85%.

$$\text{Amount of air} = \frac{0.588 \times 85}{15} = 3.332 \text{ kmol}$$

Flow rate of air $= 3.332$ kmol min^{-1}

Volume of air can be found using the ideal gas equation, $V = \frac{nRT}{P}$. When pressure is reported in atmospheres, the value of R to be used is 0.082 05 m^3 atm kmol^{-1}K^{-1}.

Volume of air at 323 K and 750 mm Hg pressure $= \dfrac{3.332 \times 0.082\,05 \times 323 \times 76}{75}$

$= 89.48\ \text{m}^3\ \text{min}^{-1}$

2J At low temperatures, nitrogen dioxide has a tendency to associate as per the following equation: $2NO_2 \rightleftharpoons N_2O_4$. 750 L of nitrogen dioxide at 685 mm Hg and 298 K are allowed to stand at constant temperature and volume, until equilibrium is reached, at which time the pressure is 500 mm Hg. Compute the partial pressure of nitrogen tetroxide in the final mixture.

Solution

Nitrogen dioxide gas in a container of known volume, exerts a certain pressure. When two molecules of nitrogen dioxide combine together to form nitrogen tetroxide, the total number of molecules becomes less and the pressure of the gas inside the container also becomes less. The total number of moles of gases in a given volume can be found using the ideal gas equation, $n = \dfrac{PV}{RT}$. When pressure is reported in atmospheres, the value of R to be used is $0.082\,05\ \text{L atm mol}^{-1}\text{K}^{-1}$.

Initial amount of nitrogen dioxide $= \dfrac{685 \times 750}{760 \times 0.082\,05 \times 298}$

$= 27.65\ \text{mol}$

Final amount of nitrogen dioxide and nitrogen tetroxide formed $= \dfrac{500 \times 750}{760 \times 0.082\,05 \times 298}$

$= 20.18\ \text{mol}$

Let x mol of nitrogen dioxide undergo association.

Amount of unassociated nitrogen dioxide $= 27.65 - x$

Amount of nitrogen tetroxide formed by association $= \dfrac{x}{2}$

Amount of gases present at the end $= (27.65 - x) + \dfrac{x}{2}$

$\therefore \quad (27.65 - x) + \dfrac{x}{2} = 20.18\ \text{mol}$

$x = 14.94\ \text{mol}$

Amount of nitrogen tetroxide formed $= 14.94 \div 2 = 7.47\ \text{mol}$

Amount of nitrogen dioxide remaining $= 27.65 - 14.94 = 12.71\ \text{mol}$

$$\text{Mole fraction of nitrogen tetroxide} = \frac{\text{number of moles of nitrogen tetroxide}}{\text{total number of moles}}$$

$$= 7.47 \div 20.18 = 0.37$$

$$\text{Partial pressure of nitrogen tetroxide} = \text{total pressure} \times \text{mole fraction}$$

$$= 500 \times 0.37$$

$$= 185 \text{ mm Hg}$$

2K **Calculate the total pressure and the composition of the vapours at 373 K for a solution containing 45% benzene, 40% toluene and 15% xylene. The vapour pressures of benzene, toluene and xylene at 373 K are 1 340, 560 and 210 mm Hg respectively**

Solution

The composition of vapours above a mixture of liquids can be found using Raoult's law which states that the partial pressure of a component in the vapour space above a solution of a liquid solute in a liquid solvent is directly proportional to the mole fraction of that component in the solution.

$$P_A \propto x_A$$

$$P_A = P_A^* x_A$$

The proportionality constant P_A^* is the vapour pressure of liquid solute, P_A is the partial pressure of the liquid solute in the vapour space above the liquid solution, and x_A is the mole fraction of the solute in the solution.

Summing up the partial pressures gives the total pressure. The mole fraction of each component in the vapour can be found by dividing the respective partial pressures by the total pressure. The composition of the vapour in mole per cent is determined by multiplying the mole fraction by 100.

The composition of the mixture is given in percentage. The valid assumption for liquids is weight per cent. As 100 g of the mixture is chosen as the basis, the weight of each of the constituents is the same as its weight per cent. The number of moles of each constituent is then found by dividing the weight by the respective molecular weight. The total number of moles is found by summing up the number of moles of each individual constituent. The mole fraction of each constituent is found by dividing the number of moles by the total number of moles. The calculations for the composition of the liquid mixture is given in Table 2.1.

Partial pressure of the constituents in the vapour phase above the liquid mixture is found using Raoult's law.

$$P_A = P_A^* x_A = 1\,340 \times 0.5 = 670.00 \text{ mm Hg}$$

$$P_B = P_B^* x_B = 560 \times 0.377 = 211.12 \text{ mm Hg}$$

$$P_C = P_C^* x_C = 210 \times 0.123 = 25.83 \text{ mm Hg}$$

$$\text{Total pressure} = 906.95 \text{ mm Hg}$$

Table 2.1 Composition of the liquid mixture

Component	Weight (g)	Molecular weight	No. of moles (weight/mol.wt)	Mole fraction (Total weight/ total no. of moles)
Benzene	45	78	0.577	0.500
Toluene	40	92	0.435	0.377
Xylene	15	106	0.141	0.123
			1.153	–

The mole fraction of each constituent is found by dividing its partial pressure by the total pressure. Table 2.2 gives the composition of the vapour phase above the liquid mixture.

Table 2.2 Composition of the vapour phase

Constituent	Partial pressure (mm Hg)	Mole fraction (total pressure/partial pressure)	Mole per cent (mole fraction × 100)
Benzene	670.00	0.738 7	73.87
Toluene	211.12	0.232 8	23.28
Xylene	25.83	0.028 5	2.85

Methods of Expressing the Composition of Mixtures

The composition of a mixture of two components may be expressed in a variety of ways. The mixture may be in any one of the three physical states—solid, liquid or gas. The composition of the mixture may be expressed on weight, mole or volume basis. If no basis is specified, the valid assumption for a solid or liquid system is weight basis; for gaseous systems, the valid assumption is volume basis.

For gaseous systems, volume basis ≡ mole basis. This is an outcome of Avagadro's hypothesis. Consider a two component system consisting of A and B.

Let W, V and n be the total weight, volume and number of moles of the system.

Let W_A, V_A and n_A be the weight, volume and number of moles of component A.

Let W_B, V_B and n_B be the weight, volume and number of moles of component B.

Let M_A and A_A be the molecular weight and atomic weight of component A.

Let M_B and A_B be the molecular weight and atomic weight of component B.

Weight fraction The weight fraction of a component is the ratio of its weight to the total weight of the system.

$$\text{Weight fraction of A} = \frac{W_A}{W}$$

$$\text{Weight fraction of B} = \frac{W_B}{W}$$

The sum of the weight fractions of A and B is 1.

Weight percentage The weight percentage of a component is the percentage ratio of its weight to the total weight of the system.

$$\text{Weight percentage (wt\%) of A} = \frac{W_A}{W} \times 100$$

$$\text{Weight percentage (wt\%) of B} = \frac{W_B}{W} \times 100$$

The sum of the weight percentages of A and B is 100.

Volume fraction The volume fraction of a component is the ratio of its volume to the total volume of the system

$$\text{Volume fraction of A} = \frac{V_A}{V}$$

$$\text{Volume fraction of B} = \frac{V_B}{V}$$

The sum of the volume fractions of A and B is 1.

Volume percentage The volume percentage of a component is the percentage ratio of its volume to the total volume of the system.

$$\text{Volume percentage (vol\%) of A} = \frac{V_A}{V} \times 100$$

$$\text{Volume percentage (vol\%) of B} = \frac{V_B}{V} \times 100$$

The sum of the volume percentages of A and B is 100.

Mole fraction The mole fraction of a component is the ratio of the number of moles of that component to the total number of moles in the system

$$\text{Mole fraction of A} = \frac{n_A}{n}$$

$$\text{Mole fraction of B} = \frac{n_B}{n}$$

The sum of the mole fractions of A and B is 1.

Mole percentage The mole percentage of a component is the percentage ratio of the number of moles of that component to the total number of moles in the system

$$\text{Mole percentage (mol\%) of A} = \frac{n_A}{n} \times 100$$

$$\text{Mole percentage (mol\%) of B} = \frac{n_B}{n} \times 100$$

The sum of the mole percentages of A and B is 100.

Atomic fraction The atomic fraction of a component is the ratio of the number of atoms of that component to the total number of atoms present in the system.

$$\text{Atomic fraction of A} = \frac{\dfrac{W_A}{A_A}}{\dfrac{W_A}{A_A} + \dfrac{W_B}{A_B}}$$

$$\text{Atomic fraction of B} = \frac{\dfrac{W_B}{A_B}}{\dfrac{W_A}{A_A} + \dfrac{W_B}{A_B}}$$

The sum of the atomic fractions of A and B is 1.

Atomic percentage The atomic percentage of a component is the percentage ratio of the number of atoms of that component to the total number of atoms in the system.

$$\text{Atomic percentage of A} = \frac{\dfrac{W_A}{A_A}}{\dfrac{W_A}{A_A} + \dfrac{W_B}{A_B}} \times 100$$

$$\text{Atomic percentage of B} = \frac{\dfrac{W_B}{A_B}}{\dfrac{W_A}{A_A} + \dfrac{W_B}{A_B}} \times 100$$

The sum of the atomic percentages of A and B is 100.

Note: Atomic fraction and atomic percentages can be calculated as given above, if components A and B are in their elemental state. If they are in the molecular state, the number of atoms of each element present in the molecules has to be taken into account.

Weight of material per unit volume of the system This mode of representation of the composition is mostly used for solutions. The weight of solute per litre of solution is denoted by gpL (grams of solute per litre of solution).

Weight of material per unit weight of a reference substance One of the components of the system may be arbitrarily chosen as a reference material and the composition of the system may be indicated by stating the weight of each component associated with unit weight of the reference material.

Parts per thousand The unit, parts per thousand (ppt), denotes the number of parts of a component per one thousand parts of the system.

Parts per million The unit, parts per million (ppm), denotes the number of parts of a component per one million parts of the system.

$$1000 \text{ ppm} \equiv 1 \text{ ppt}$$

Parts per billion The unit, parts per billion (ppb), denotes the number of parts of a component per one billion parts of the system.

$$1000 \text{ ppb} \equiv 1 \text{ ppm}$$

Normality (N) Normality denotes the number of equivalents of a solute in one litre of a solution.

Molarity (M) Molarity denotes the number of moles of a solute in one litre of a solution.

Molality Molality denotes the number of moles of a solute in one kilogram of a solvent.

Note: The last three modes are used only for liquid solutions. For dilute aqueous solutions, that is for solutions whose density is almost equal to that of water: ppt $\equiv$ gpL; ppm $\equiv$ mgpL; ppb $\equiv$ mgpL.

ILLUSTRATIONS

2L Brackish water contains 16 000 ppm sodium chloride. Express the concentration in weight per cent.

Solution

Weight of sodium chloride in brackish water = 16 000 parts per 1 000 000 parts

$$\text{Weight per cent of sodium chloride in brackish water} = \frac{16\,000}{1\,000\,000} \times 100 = 1.6\%$$

2M Commercial sulphuric acid is 98% sulphuric acid and 2% water. What is the mole ratio of sulphuric acid to water

Solution

Basis: 100 g of acid (The substance is a liquid. The valid assumption for liquids is weight per cent. If we choose 100 g as the basis, weight of each component will be equal to its weight per cent.)

The calculation is given in Table 2.3.

Table 2.3 Composition of the mixture

Component	Weight (g)	Molecular weight	Moles (weight/mol. weight)
H_2SO_4	98	98	1.0
H_2O	2	18	0.111

Mole ratio of sulphuric acid to water $= 1:0.111$

$= 9:1$ (Ratio should be a whole number)

2N A sample of urea contains 45% of nitrogen. Find its purity.

Solution

Basis: 100 g of urea (The substance is a solid. The valid assumption for solids is weight per cent. If we choose 100 g as the basis, weight of each component will be equal to its weight per cent.)

Weight of nitrogen in the sample taken $= 45$ g

The molecular weight of urea (H_2NCONH_2) is 60. One molecule of urea contains two atoms of nitrogen or 60 g of urea contains 28 g of nitrogen. In otherwords, 28 g of nitrogen are equivalent to 60 g of urea.

Actual urea content in the sample taken $= 45 \times \dfrac{60}{28} = 96.42$ g

Hence purity of the sample $= 96.42\%$

2O A sample of medium grade bituminous coal on analysis gave the following result: sulphur 2%, nitrogen 1%, oxygen 6%, ash 11% and water 3%. The residum of carbon and hydrogen is in the mole ratio C:H = 9:1. Calculate the weight fraction composition of coal with ash and moisture omitted.

Solution

Basis: 100 g coal (The substance is a solid. The valid assumption for solids is weight per cent. If we choose 100 g as the basis, weight of each component will be equal to its weight per cent.)

Carbon and hydrogen are in the mole ratio of 9:1. This should be converted into weight ratio, as we have chosen 100 g of coal (weight) as the basis. This has to be done by multiplying the number of moles of each component by its atomic weight, if it is an element, or by its molecular weight if it is a molecule.

$$\begin{aligned} C:H &= 9:1 \text{ (mole ratio)} \\ &= 9 \times 12 : 1 \times 2 \text{ (weight ratio)} \\ &= 108:2 \\ &= 54:1 \end{aligned}$$

(Carbon is an element and hydrogen is a molecule. 12 is the atomic weight of carbon; 2 is the molecular weight of hydrogen.)

$\therefore$ 55 parts of carbon and hydrogen contain 54 parts of carbon and 1 part of hydrogen.

It is given that the coal contains 2% sulphur, 1% nitrogen, 6% oxygen, 11% ash and 3% water, and the residum is carbon and hydrogen. As all percentages are by weight, the amount of carbon and hydrogen in 100 g coal is 77 g.

$$\text{Weight of carbon} = 77 \times \frac{54}{55} = 75.6 \text{ g}$$

$$\text{Weight of hydrogen} = 77 - 75.6 = 1.4 \text{ g}$$

Composition of coal on ash- and moisture-free basis is given in Table 2.4. Weight fraction of each component is found by dividing its weight by total weight.

Table 2.4 Composition of coal on ash and moisture-free basis

Component	Weight (g)	Weight fraction (weight/total weight)
S	2	0.023 3
N_2	1	0.011 7
O_2	6	0.069 9
C	75.6	0.879
H_2	1.4	0.016 1
	86.0	

2P What will be the percentage of Na_2O content in lye containing 78% NaOH?

Solution

Basis: 100 g of NaOH aqueous solution (The substance is a liquid. The valid assumption for liquids is weight per cent. If we choose 100 g as the basis, the weight of each component will be equal to its weight per cent.)

Weight of NaOH in the aqueous solution = 78 g

It is customary to express the concentration of alkali in terms of the alkaline oxides.

$$\underset{80}{2NaOH} \rightarrow \underset{62}{Na_2O} + H_2O$$

80 g of NaOH is equivalent to 62 g of Na_2O.

$$\text{NaOH content expressed as } Na_2O = \frac{78}{80} \times 62 = 60.45\%$$

2Q A solution of sodium chloride in water contains 230 gpL of NaCl at 293 K. The density of the solution at that temperature is 1.148 g cm^{-3}. Calculate the following: composition by weight per cent, composition by mole per cent, composition by atomic per cent and volume per cent of water. The density of water at 298 K is 0.998 g cm^{-3}.

Solution

Basis: 1 L of solution weighing 1148 g (The composition of the solution is given in grams per litre. So the basis has been chosen as one litre of solution and its weight has been found by multiplying volume by density.)

Weight of solution = 1 148 g

Weight of sodium chloride = 230 g

Weight of water = 1 148 − 230 = 918 g

The weight per cent of each component is the percentage ratio of its weight to the total weight of the solution. The number of moles of each constituent is calculated by dividing the weight by the respective molecular weight. The total number of moles is found by summing up the individual moles of the constituents. The mole per cent of each component is the percentage ratio of its number of moles to the total number of moles. Table 2.5 explains the actual calculation.
The volume of water is calculated by dividing the weight of water by the density of water. Volume per cent of water is the percentage ratio of its volume to the total volume, that is, volume of the solution.

$$\text{Volume of water} = \frac{918}{0.998} = 920 \text{ cm}^3$$

Volume per cent of water = 92%

Table 2.5 Composition of the solution

Constituent	Weight (g)	Weight per cent	Molecular weight	No. of moles	Mole per cent
NaCl	230	20.03	58.46	3.93	7.15
H_2O	918	79.97	18	51	92.85
	1148			54.93	

One mole of an element contains Avagadro number (N) of atoms. Based on this concept, the number of atoms of each element is calculated. The total number of atoms is found by summing up the atoms of the individual constituents. The atomic per cent of each component is the percentage ratio of its atoms to the total number of atoms. Table 2.6 explains the actual calculation.

Table 2.6 Composition of the solution

Constituent	No. of atoms	Atom per cent
Na	3.93 N	2.45
Cl_2	3.93 N	2.45
H_2	102.00 N	63.80
O_2	51.00 N	31.90
	159.86 N	

2R **A sample of aqueous triethanolamine (TEA) contains 47% TEA by volume. If the density of pure TEA and water are 1.125 g cm^{-3} and 1.00 g cm^{-3} respectively, find the weight per cent TEA in the solution.**

Solution

Basis: 100 cm^3 of solution (If we choose 100 cm^3 as the basis, the volume of each component will be equal to its volume per cent.)

Volume is converted to weight by multiplying the volume of each constituent by its density. The weight per cent of each component is the percentage ratio of its weight to the total weight, that is, weight of solution. Table 2.7 explains the actual calculation.

Table 2.7 Composition of the solution

Constituent	Volume (cm^3)	Density (g cm^{-3})	Weight (g)	Weight per cent
TEA	47	1.125	52.875	49.94
H_2O	53	1.00	53.00	50.06
			105.875	

2S A solution of sodium hydroxide in water contains 20% NaOH by weight at 333 K. Density of the solution at that temperature is 1.2 g cm^{-3}. Find the concentration of the solution in gpL, normality, molarity and molality.

Solution

Basis: 1 L of solution (It is required to find the concentration of the solution in gpL. So this is the apt basis.)

The weight of the solution is found by multiplying the volume of the solution by the density of the solution. Then the weight of solute and solvent are found from the respective weight per cent given in the problem.

Weight of 1 L of solution $= 1000 \times 1.2 = 1\,200$ g

Weight of the solute, sodium hydroxide $= 1200 \times \frac{20}{100} = 240$ g

Weight of water $= 1\,200 - 240 = 960$ g

Concentration in gpL of the solution or the weight of NaOH in one litre of solution = 240 g. The normality of the solution is the number of equivalents of solute in one litre of solution. The molarity of the solution is the number of moles of solute in one litre of solution. For sodium hydroxide, molecular weight and equivalent weight are 40.

Normality or molarity of the solution $= \frac{240}{40} = 6$ N

The molality of the solution is the number of moles of solute in 1000 g of solvent. 960 g of solvent contain 6 mol of solute.

Amount of solute in 1000 g of solvent $= \frac{6}{960} \times 1000 = 6.25$ g

Molality of the solution $= 6.25$ m

2T 1000 m^3 of a mixture of hydrogen, nitrogen and carbon dioxide at 423 K was found to have the following ratio for the partial pressures of the gases. $P_{H_2} : P_{N_2} : P_{CO_2} = 1:4:3$. If the total pressure is 2 atm, calculate (i) the mole fraction of each gas, (ii) the weight per cent of these gases, (iii) the average molecular weight, and (iv) the weight of carbon dioxide present in kilograms.

Solution

Basis: 1 mol of the gaseous mixture (It is required to compute mole fraction and average molecular weight. The mole fraction of each component is the number of its moles if the total number of moles is one. The average molecular weight is the weight of one mole of the mixture of gases. In view of the above, the basis chosen is the apt one.)

By Dalton's Law of partial pressures, ratio of partial pressures ≡ ratio of number of moles

$H_2 : N_2 : CO_2 = 1:4:3$ (number of moles)

Total number of moles = 8

Mole fraction of each component is found by dividing its number of moles by total number of moles.

$$\text{Mole fraction of hydrogen} = \frac{1}{8} = 0.125$$

$$\text{Mole fraction of nitrogen} = \frac{4}{8} = 0.500$$

$$\text{Mole fraction of carbon dioxide} = \frac{3}{8} = 0.375$$

The weight of each component is calculated by multiplying the mole fraction by the respective molecular weight. The total weight is found by summing up the individual weights of the constituents. Incidentally, the total weight is also the weight of one mole of the gaseous mixture, which is its average molecular weight. The weight per cent of each component is the percentage ratio of its weight to the total weight. Table 2.8 explains the actual calculation.

Table 2.8 Composition of the gaseous mixture

Constituent	Mole fraction	Molecular weight	Weight	Weight per cent
H_2	0.125	2.016	0.252	0.82
N_2	0.50	28	14.00	45.53
CO_2	0.375	44	16.500	53.65
			30.752	

Average molecular weight = 30.752

To find the weight of carbon dioxide in 1000 m^3 of the gaseous mixture, the number of moles of the gaseous mixture is found using the formula, $n = \frac{PV}{RT}$. When pressure is reported in atmosphere, volume in m^3, the value of R to be used is 0.082 05 m^3 atm kmol^{-1}K^{-1}.

$$\text{Total number of moles in the gaseous mixture} = \frac{2 \times 1000}{0.082\,05 \times 423} = 57.62 \text{ kmol}$$

Amount of carbon dioxide = total number of moles × mole fraction

$$= 57.62 \times 0.375 = 21.607\,5 \text{ kmol}$$

Weight of carbon dioxide = Number of moles × molecular weight

$$= 21.607\,5 \times 44 = 950.73 \text{ kg}$$

2U A gaseous mixture has the following composition by volume—methane 10%, ethane 30% and hydrogen 60%—and flows at the rate of 1000 L s^{-1} at 303 K and 2000 mm Hg gauge pressure. Compute: (i) the mole fraction of each component, (ii) the concentration of each component in mol L^{-1}, (iii) the partial pressure of each component, (iv) the average molecular weight, and (v) the mass flow rate in kg h^{-1}.

Solution

Basis: 1 mol of the gaseous mixture (It is required to compute the mole fraction and average molecular weight. The mole fraction of each component is the number of its moles if the number of total moles is one. The average molecular weight is the weight of one mole of the mixture of gases. In view of the above, the basis chosen is the apt one.)

We know that for gases, volume per cent ≡ mole per cent. So the composition of the gases in volume per cent is also their mole per cent. Mole per cent can be converted to mole fraction by dividing mole per cent by 100, since per cent is out of 100 and fraction is out of one.

Mole fraction of methane = 0.1

Mole fraction of ethane = 0.3

Mole fraction of hydrogen = 0.6

The weight of each constituent is calculated by multiplying the mole fraction by its molecular weight. The total weight is found by summing up the individual weights of the constituents. Incidentally, the total weight is also the weight of one mole of the gaseous mixture, which is its average molecular weight. Table 2.9 explains the actual calculation.

Table 2.9 Composition of the gaseous mixture

Constituent	Mole fraction	Molecular weight	Weight (g)
CH_4	0.1	16	1.6
C_2H_6	0.3	30	9.0
H_2	0.6	2.016	1.21
			11.81

Average molecular weight = 11.81

Partial pressure of a component = Total pressure × mole fraction

Total pressure = 2 000 mm Hg gauge = 2 000 + 760 = 2 760 mm Hg absolute

Partial pressure of methane = 2 760 × 0.1 = 276 mm Hg

Partial pressure of ethane = 2 760 × 0.3 = 828 mm Hg

Partial pressure of hydrogen = 2 760 × 0.6 = 1 656 mm Hg

Mass flow rate $= \text{volumetric flow rate} \times \text{density}$.

Density $= \dfrac{PM}{RT} = \dfrac{2\,760 \times 11.81}{760 \times 0.082\,05 \times 303} = 1.725\ \text{kg m}^{-3}$

Volumetric flow rate $= 1000\ \text{L s}^{-1} = 1\ \text{m}^3\ \text{s}^{-1} = 3\,600\ \text{m}^3\ \text{h}^{-1}$

Mass flow rate $= 3\,600\ \text{m}^3\ \text{h}^{-1} \times 1.725\ \text{kg m}^{-3} = 6\,210\ \text{kg h}^{-1}$

To find the concentration of each component in one litre of the gaseous mixture, the number of moles in 1 L of the gaseous mixture is found using the formula, $n = \dfrac{PV}{RT}$. When pressure is reported in atmospheres, and volume in litres, the value of R to be used is 0.082 05 L atm $\text{mol}^{-1}\text{K}^{-1}$.

Total number of moles per litre of gaseous mixture $= \dfrac{2\,760}{760} \times \dfrac{1}{0.082\,05 \times 303}$

$= 0.146\ \text{mol}$

Concentration of methane $= \text{total number of moles} \times \text{mole fraction}$

$= 0.146 \times 0.1 = 0.014\,6\ \text{mol L}^{-1}$

Concentration of ethane $= 0.146 \times 0.3 = 0.043\,8\ \text{mol L}^{-1}$

Concentration of hydrogen $= 0.146 \times 0.6 = 0.087\,6\ \text{mol L}^{-1}$

2V A gaseous mixture consists of three components, A, B and C. Analysis of the mixture showed it contains 40mol% A, 18.75mass% B and 20mol% C. The molecular weight of A and C are 40 and 50 respectively. Find the molecular weight of B and the average molecular weight of the gas mixture.

Solution

Basis: 1 mol of the gaseous mixture

B constitutes 18.75mass% of the gaseous mixture and the balance, 81.25mass% of gas mixture, is A and C. The mole per cent and molecular weight of A and C are known; the mole fraction of A and C can be computed by dividing their mole per cent by 100. The weight of A and C can be found by multiplying the mole fraction by the respective molecular weight.

81.25mass% of gas mixture $= (0.4 \times 40) + (0.2 \times 50)$

$= 26\ \text{g}$

100 mass% of the gas $= \dfrac{26}{0.812\,5} = 32\ \text{g}$

Thus the average molecular weight of the gas mixture = 32.

Mole fraction of B in the gaseous mixture	= 1.0 – sum of the mole fractions of A and B
	= 1.0 – 0.6 = 0.4
Weight of B in one mole of the gaseous mixture	= average molecular weight – total weight of A and B
	= (32 – 26) = 6 g
Weight of 0.4 mol of B	= 6 g
Molecular weight of B	$= \frac{6}{0.4} = 15$

2W The average molecular weight of a flue gas sample containing carbon dioxide, oxygen and nitrogen is calculated by an engineer and a chemist. The chemist used the correct molecular weight of nitrogen and determined the average molecular weight of the flue gas to be 30.6. The engineer used the atomic weight of nitrogen as the molecular weight and calculated the average molecular weight to be 19.4. Compute the composition of the flue gas.

Solution

Let the mole fractions of carbon dioxide, oxygen and nitrogen be X_{CO_2}, X_{O_2} and X_{N_2}. We have three unknowns and to get them we must have three equations. The equations can be framed based on the following concepts:

- Sum of the mole fractions is equal to one
- Sum of the multiples of the mole fraction of each constituent and its molecular weight is equal to average molecular weight

$$X_{CO_2} + X_{O_2} + X_{N_2} = 1 \quad (1)$$

$$44X_{CO_2} + 32X_{O_2} + 28X_{N_2} = 30.6 \quad (2)$$

$$44X_{CO_2} + 32X_{O_2} + 14X_{N_2} = 19.4 \quad (3)$$

From equations, (1), (2) and (3) we get

$X_{N_2} = 0.8$

$X_{O_2} = 0.05$

$X_{CO_2} = 0.15$

The mole per cent of a gas can be obtained by multiplying the mole fraction by 100. For a gaseous system the composition should be expressed in mole per cent which is also equal to volume per cent. The composition of the flue gas is given in Table 2.10.

Table 2.10 Composition of the flue gas

Constituent	Volume per cent or Mole per cent
CO_2	15
O_2	5
N_2	80

Problems for Practice

2.1 Find the volume in cubic metres of 40 kg of butane stored at 323 K and 600 bar.

2.2 The strength of phosphoric acid as phosphorous pentoxide is 34%. Find the actual concentration.

2.3 The available nitrogen content of an ammonium nitrate sample is 34%. Report its purity.

2.4 How many moles of hydrogen will occupy 1000 m^3 at a pressure of 112×10^3 N m^{-2} and a temperature of 400 K?

2.5 Equal masses of carbon monoxide and nitrogen are mixed at 300 K. If the total pressure is 400 kPa, find the partial pressure of carbon monoxide.

2.6 Calculate the weight of 10 m^3 of water vapour at a pressure of 20 mm Hg and 300 K temperature.

2.7 It is required to compress 10 kg of nitrogen at a temperature of 300 K to a volume of 1 m^3. Calculate the pressure required.

2.8 Calculate the volume occupied by 50 kg of chlorine gas at a pressure of 100 kPa and temperature 298 K.

2.9 5 kg of oxygen in a closed container of volume 1 m^3 is heated, allowing the pressure not to exceed 800 kPa. Calculate the maximum temperature which can be attained by the gas.

2.10 Calculate the average molecular weight of flue gas with the following composition by volume: carbon dioxide 13%, oxygen 8% and nitrogen 79%.

2.11 A mixture of gases has an average molecular weight of 43. It contains 40% carbon dioxide, 20% carbon monoxide, 5% oxygen and the rest is an unknown gas. What is the molecular weight of the unknown gas?

2.12 A mixture of methane and ethane has an average molecular weight of 24.4. Find the mole per cent of methane in the mixture.

2.13 A solution of caustic soda contains 30% sodium hydroxide by weight. The density of the solution is 1.196 g cm^{-3}. Find the normality, molarity and molality of the solution.

2.14 A sample of wine contains 20% alcohol on volume basis. Find the weight per cent of alcohol in the wine. The density of alcohol and alcohol-free liquid are 0.79 g cm^{-3} and 1 g cm^{-3} respectively.

2.15 An aqueous solution of acetic acid of concentration 40% by weight has a density of 1.04 g cm^3. Find the molality, molarity and normality.

2.16 100 g of nitric acid and 100 g of sulphuric acid are filled in two bottles. Which bottle contains more atoms and how much more?

2.17 How many kilograms of hydrogen chloride are contained in one metric tonne of 36.5% hydrochloric acid with a specific gravity of 1.18?

2.18 An aqueous solution contains 2.29mol% magnesium chloride. The specific gravity of the solution at 293 K is 1.9. Calculate the concentration of the solution in terms of kilograms of solute per cubic metre of solution.

2.19 It is desired to prepare a 10 molar solution of sodium hydroxide in water at 303 K. Calculate the weight of sodium hydroxide that should be placed in a 500 ml standard flask in order that the desired concentration will be obtained by subsequently filling the flask with water. The density of the solution at 303 K is 1.32 g cm^{-3}.

2.20 The density of a 20wt% sodium chloride solution in water is 1.155 g cm^{-3} at 277 K. (i) Calculate the mole fraction of sodium chloride, normality and molarity of the solution. (ii) If 500 cm^3 of water is added to 1000 cm^3 of the solution, what will be the molality of the solution and the volume percentage of water if the process takes place at 277 K? Density of water at that temperature is 1 g cm^{-3}.

2.21 A solution of zinc bromide in water contains 130 grams of ZnBr per 100 ml of solution. The density of the solution is 2.003 g cm^{-3} at 293 K. Compute: (i) the composition in weight per cent, (ii) the composition in mole per cent, (iii) the composition in atomic per cent, (iv) the molality and (v) the molarity.

2.22 A liquid contains 40% pentane, 25% hexane and 35% heptane. Calculate the mole per cent of each component and the average molecular weight. (Molecular weight of pentane, hexane and heptane are 72, 86 and 100 respectively.)

2.23 A gaseous mixture contains 40 kg benzene, 40 kg toluene and 30 kg xylene. Calculate the mole per cent of each component and the average molecular weight. (Molecular weight of benzene, toluene and xylene are 78, 92 and 106 respectively.)

2.24 Calculate the average molecular weight of a gaseous mixture containing 50 g each of methane, hydrogen and nitrogen.

2.25 When heated to 373 K and 720 mm Hg, 25 kg of nitrogen tetroxide gas occupies a volume of 16.42 m^3. Calculate the per cent dissociation of nitrogen tetroxide to nitrogen dioxide.

2.26 When nitrogen tetroxide is heated to 673 K, 60% of it dissociates. If 50 kg of nitrogen tetroxide is heated to 673 K and 740 Torr pressure, calculate the volume occupied by the resulting gaseous mixture.

2.27 Carbon dioxide at high temperatures dissociates according to the following equation: $2CO_2 \rightleftharpoons 2CO + O_2$. At 3 273 K, 40% carbon dioxide gets dissociated. If 5 litres of carbon dioxide at 293 K and 1 atm is heated to 3273 K at constant pressure, find the volume of the resulting gases.

2.28 A mixture of gases has the following composition: nitrogen 65%, oxygen 18%, ammonia 9% and the balance water vapour. Compute (i) the composition in weight per cent, and (ii) the volume in cubic metres of 10 kg of this mixture at 873 K and 10 atm gauge pressure.

2.29 A mixture of nitrogen and carbon dioxide at 303 K and 2 atm gauge pressure has an average molecular weight of 31. Compute (i) the volume fraction of carbon dioxide, (ii) the partial pressure of nitrogen, and (iii) the number of atoms in 5 kg of this mixture.

2.30 A gas has a composition by volume as follows: methane 78%, ethane 12% and propane 10%. Calculate (i) the composition by weight per cent, (ii) the average molecular weight, (iii) the density at 740 mm Hg pressure and 423 K, and (iv) the specific gravity.

2.31 A mixture of gases has the following composition by weight: chlorine 55%, bromine 40% and the balance is oxygen. Using the ideal gas law, compute (i) the composition in mole per cent, (ii) the density of the gas in kg m^{-3} at 298 K and 800 Torr, and (iii) the specific gravity.

2.32 A natural gas containing 3% hydrogen sulphide, 4% butane, 3% propane, 3% ethane, 82% methane, 2% helium and 3% nitrogen is piped from the well at an absolute pressure of 4 bar and 300 K. Calculate (i) the partial pressure of each component, (ii) the pure component volume of the components, and (iii) the density of the gas at above conditions.

2.33 A solution contains 50% benzene, 30% toulene and the rest *o*-xylene by weight at 373 K. The vapours are in contact with the solution. Calculate the total pressure and mole per cent composition of the liquid and vapour. State and explain your assumptions. Vapour pressure of the components at 373 K is: benzene 1.76 bar, toluene 0.74 bar and *o*-xylene 0.28 bar.

2.34 Pure nitrogen at 360 K and 1.2 bar is flowing through a pipe. Hydrogen is to be added to this stream in such an amount as to produce a 1:3 nitrogen to hydrogen mixture suitable for ammonia synthesis. If hydrogen is at 300 K and 2.5 bar, calculate the flow rates of nitrogen and hydrogen streams in cubic metres per minute for the production of 100 kmol of ammonia per hour.

3 Material Balance in Non-Reaction Systems

Material balance of a process is the exact accounting of all the materials that enter, leave, accumulate or deplete in the course of a given time interval of operation. The law of conservation of mass stipulates that for any process

$$\text{Input} - \text{accumulation} = \text{output}$$

In any flow or batch process, a definite mass of material is charged to the equipment. If there is no accumulation, or accumulation is constant,

$$\text{Input} = \text{output}$$

This happens in the case of all steady state processes. Consider the overhead water tank. If the rate at which the tank is filled and the rate at which water is drained from the tank are equal, there will be no accumulation. On the other hand, if the rate at which the tank is filled is faster than the rate at which water is drained from the tank, there will be accumulation.

Material balance problems can be classified as

- Material balance of processes involving no chemical reaction
- Material balance of processes involving chemical reaction.

The common scheme outlined below may be followed for solving material balance problems:

1. Draw a block flow diagram representing each process by a single block.
2. Transfer all the quantitative information such as rate of different streams entering, and leaving, and their compositions, to the flow diagram.
3. Identify a component which has one input and only one output. Such a component is the key component or *tie material*.
4. Select a suitable basis for calculation. For a flow process, some unit of time may be the suitable basis. For a batch process, quantity of one of the components present in the system may be the basis. For a semi-flow or semi-batch process, either time or the quantity of one of the components of the system may be taken as the basis.
5. Write the overall material balance equations and material balance equations for each individual component so that the number of unknowns and the number of equations match each other.
6. Solve the equations to get the required solution.

Example

10 kg of salt is mixed with 100 kg of water to get 110 kg of salt solution. The process is known as dissolution. No chemical reaction is involved. The method of representing the data provided in the form of a block diagram (Fig. 3.1) and framing of material balance equations are outlined below.

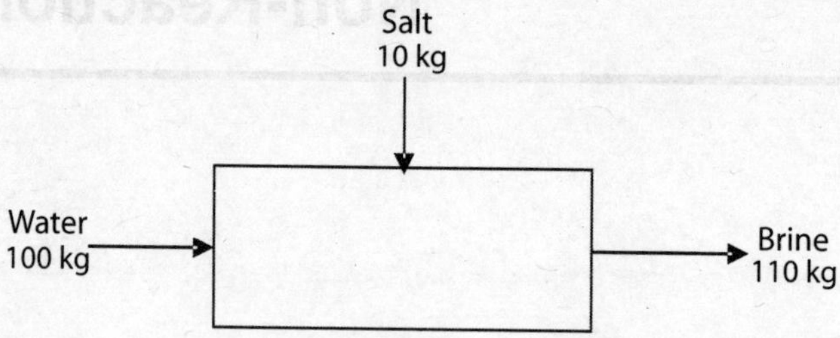

Fig. 3.1 Schematic representation of dissolution operation

Overall balance 100 kg water + 10 kg salt = 110 kg solution

Solute balance 0 kg solute in solvent + 10 kg solute = 10 kg solute in solution

Solvent balance 100 kg solvent + 0 kg solvent in solute = 100 kg solvent in solution

Note:

1. Usually the basis stated in the problem may be used.
2. If it is not convenient, a more suitable basis may be assumed.
3. The problem should be solved using the same units throughout.

ILLUSTRATIONS

3A A material is to be dried in hot air and the moisture content reduced from 20% to 1%. Fresh air contains 0.02 kg of water per kilogram of dry air. Find the volume of fresh air required, if 1000 kg of dried material is to be produced. The exit air contains 0.09 kg of water per kilogram of dry air. Air enters at 300 K.

Solution

Basis: 1000 kg of dried material

Figure 3.2 represents the data provided in the form of a block diagram. The dried material is the product from the process and it contains 1% moisture.

$$\text{Moisture in the product} = 1000 \times \frac{1}{100} = 10\text{ kg}$$

$$\text{Moisture-free solids} = 1000 - 10 = 990\text{ kg}$$

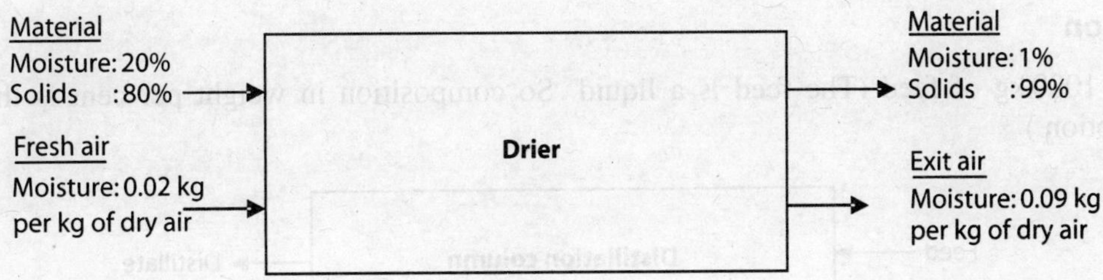

Fig. 3.2 Drier

Incoming feed contains 20% moisture and 80% moisture-free solids. The source for the moisture-free solids in the product is the feed. Moisture-free solids entering the drier is 990 kg. This forms 80% of the feed.

Moisture in the feed $= 990 \times \dfrac{20}{80} = 247.5\,\text{kg}$

Moisture in the product $= 10\,\text{kg}$

Moisture removed in the drier $= 237.5\,\text{kg}$

The material is dried by hot air. Hot fresh air fed to the drier contains 0.02 kg of moisture per kilogram of dry air. The hot air picks up the moisture lost by the solid being dried. The moisture content of air leaving the drier is 0.09 kg per kilogram of dry air.

Moisture picked up per kilogram of dry air $= 0.09 - 0.02 = 0.07\,\text{kg}$

Fresh dry air required $= \dfrac{237.5}{0.07}$

$$= 3\,392.86\,\text{kg} = \frac{3\,392.86}{29} = 116.99\,\text{kmol}$$

Hot fresh air fed to the drier contains 0.02 kg of moisture per kilogram of dry air.

Moisture accompanying fresh dry air $= 3\,392.86 \times 0.02 = 67.86\,\text{kg}$

$$= \frac{67.86}{18} = 3.77\,\text{kmol}$$

Fresh air entering the drier $= 116.99 + 3.77 = 120.76\,\text{kmol}$

Volume of fresh air at 300 K and atmospheric pressure $= \dfrac{nRT}{P}$

$$= \frac{120.76 \times 0.082\,05 \times 300}{1} = 2\,972.5\,\text{m}^3$$

3B A plant distills 1000 kg of a 20% solution of acetone in water. 50 kg of pure acetone is lost to the water residue. The distillate contains 90% acetone and 10% water. Calculate the weight of distillate and residue.

Solution

Basis: 1000 kg of feed (The feed is a liquid. So composition in weight per cent is the valid assumption.)

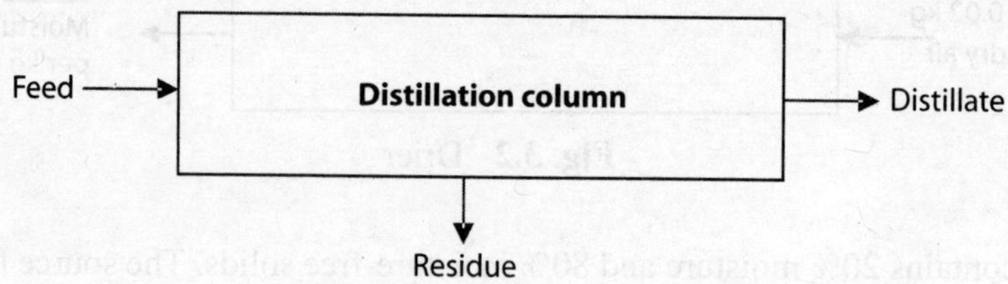

Fig. 3.3 Distillation column

Figure 3.3 represents the data provided in the form of a block diagram.

The distillation column splits the feed (F) into distiliate (D) and residue (R).

Overall material balance: Feed $=$ distillate $+$ residue

$$F = D + R \tag{1}$$

The feed is a mixture of acetone and water. The amount of acetone in the feed is 20% by weight.

Amount of acetone in feed $= 1000 \times \dfrac{20}{100} = 200\,\text{kg}$

The distillate contains 90% by weight (0.9 weight fraction) of acetone. The residue contains 50 kg acetone. Acetone has one input (feed) and two outputs (distillate and residue)

Acetone balance: Acetone in feed $=$ acetone in distillate $+$ acetone in residue

$$200 = 0.9D + 50 \tag{2}$$

We have two unknowns and two equations for getting the value of the unknowns. Solving equations (1) and (2) we get:

Weight of distillate $= 2\,167\,\text{kg}$

Weight of residue $= 7\,833\,\text{kg}$

3C In the manufacture of hydrochloric acid a gas containing 25% HCl gas and the rest air is obtained. This gas is passed through an absorption column in which 98% HCl is removed. The gas enters the system at 322 K and 743 mm Hg and leaves at 300 K and 738 mm Hg. Calculate (i) the volume of gases leaving per 100 L of gas entering, (ii) the composition of gases leaving the system, and (iii) the weight of hydrochloric acid removed per 100 L of gases entering.

Solution

Basis: 100 L of gases entering (Basis given in the problem is chosen.)

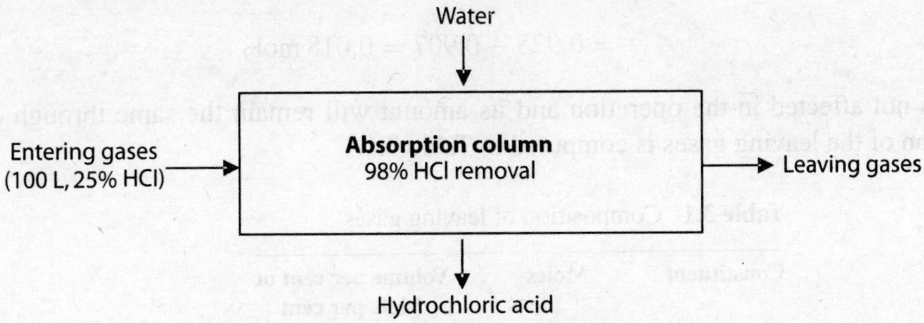

Fig. 3.4 Absorption column

Figure 3.4 represents the data provided in the form of a block diagram.

As a gaseous system is involved, the problem should be solved on molar basis.

The number of moles of the gases in a known volume can be found using ideal gas equation. $n = \dfrac{PV}{RT}$. When the pressure is reported in atmospheres, the value of R to be used is 0.082 05 L atm mol^{-1}K^{-1}.

Amount of entering gases $= \dfrac{743 \times 100}{760 \times 0.08205 \times 322} = 3.7$ mol

The entering gases contain 25% HCl.

Amount of hydrochloric acid in the entering gases $= \dfrac{3.7 \times 25}{100} = 0.925$ mol

Amount of inert air in the entering gases $= 3.7 - 0.925 = 2.775$ mol

In the absorber, 98% of entering hydrochloric acid is removed by contact with water.

Amount of hydrochloric acid removed $= \dfrac{0.925 \times 98}{100} = 0.907$ mol

Weight of hydrochloric acid removed $= 0.907 \times 36.47 = 33.06$ g

Amount of leaving gases = number of moles of entering gases – number of moles of HCl removed

$= 3.700 - 0.907 = 2.793$ mol

Volume of leaving gases at 300 K and 738 mm Hg $= \dfrac{nRT}{P}$

$$= \frac{2.793 \times 0.08205 \times 300 \times 760}{738} = 70.80 \text{ L}$$

Amount of hydrochloric acid in the leaving gases = number of moles of HCl in the entering gases – number of moles of HCl removed

$$= 0.925 - 0.907 = 0.018 \text{ mol}$$

Inert air is not affected in the operation and its amount will remain the same through out. The composition of the leaving gases is computed in Table 3.1.

Table 3.1 Composition of leaving gases

Constituent	Moles	Volume per cent or Mole per cent
Air	2.775	99.36
HCl	0.018	0.64
	2.793	

For gases, the valid assumption for expressing composition is volume or mole per cent which are equivalent.

3D Absorption of sulphur dioxide is carried out in a packed tower. Gas containing 14.8% sulphur dioxide and the rest inert gases enters the tower while the leaving gases contain 1% sulphur dioxide. Water flows at the rate of 1 $m^3 min^{-1}$. The tower handles 7 380 $m^3 h^{-1}$ of gas at 303 K and 1 bar. Find the sulphur dioxide concentration of the effluent from the tower. Also find the volume of the gases leaving the tower at 0.95 bar and 293 K.

Solution

Basis: The process is a flow process. Operation for a particular period of time is the apt basis. The gas flow rate is given on hourly basis and the water flow rate is given on minute basis. It is

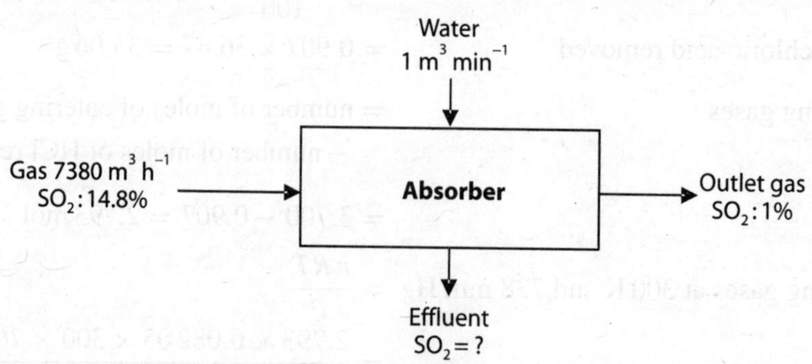

Fig. 3.5 Absorption column

always better to select the longer time as the basis. One hour operation is the best basis for this problem.

Figure 3.5 represents the data provided in the form of a block diagram.

The operation involves a gaseous system and should be solved on molar basis. The valid assumption for the composition is mole per cent. Gas is fed to the tower at the rate of 7 380 m^3 h^{-1} of gas at 303 K and 1 bar.

Amount of entering gases $= \dfrac{PV}{RT}$

$$= \frac{1 \times 7\,380}{1.013 \times 0.082\,05 \times 303} = 293.04 \text{ mol}$$

The entering gas contains 14.8mol% sulphur dioxide gas and the rest inert gases.

Amount of sulphur dioxide in the entering gases $= \dfrac{293.04 \times 14.8}{100} = 43.37$ kmol

Amount of inert gases $= 293.04 - 43.37 = 249.67$ kmol

The inert gases will be unaffected by the operation and their amount will remain the same throughout. Outgoing gases contain 1% of sulphur dioxide and 99% of inert gases.

Amount of sulphur dioxide in the outgoing gases $= \dfrac{249.67 \times 1}{99} = 2.52$ kmol

Amount of sulphur dioxide that is absorbed = amount of sulphur dioxide entering − amount of sulphur dioxide leaving

$$= 43.37 - 2.52 = 40.85 \text{ kmol}$$

Weight of sulphur dioxide absorbed $= 40.85 \times 64 = 2\,614.4$ kg

Water is fed to the tower at the rate of 1 $m^3 min^{-1}$. Therefore, water fed in one hour is 60 m^3.

Weight of water fed = 60 000 kg

Weight of effluent = weight of water + weight of sulphur dioxide absorbed

$$= 60\,000 + 2\,614.4 = 62\,614.4 \text{ kg}$$

Weight per cent of sulphur dioxide in the effluent $= \dfrac{2\,614.4 \times 100}{62\,614.4} = 4.18$wt%

For liquids, the valid assumption for expressing composition is weight per cent.

Amount of leaving gases = amount of inert gas + amount of unabsorbed sulphur dioxide

$$= 249.67 + 2.52 = 252.19 \text{ kmol}$$

Volume of leaving gases at 293 K and 0.95 bar $= \dfrac{nRT}{P}$

$$= \frac{252.19 \times 0.082\,05 \times 293 \times 1.013}{0.95} = 6\,464.87\ \text{m}^3$$

3E Combustion gases with the following molal composition are passed into an evaporator at 473 K and 743 mm Hg: nitrogen 79.2%, oxygen 7.2% and carbon dioxide 13.6%. Water is added to the stream as vapour and the gases leave at 358 K and 740 mm Hg with the following composition: nitrogen 48.3%, oxygen 4.4%, water vapour 39% and the balance is carbon dioxide. Calculate the volume of gases leaving the evaporator per 100 L of gases entering, and the weight of steam added to the gases.

Solution

Basis: 100 L of entering gases

Amount of entering gases $= \dfrac{PV}{RT} = \dfrac{743 \times 100}{760 \times 0.082\,05 \times 473}$

$$= 2.52\ \text{mol}$$

In the evaporator, steam is added to the entering gaseous mixture. Outgoing gases are found to contain 39% steam and the rest of the 61% is the entering gases, that is, 61% of the leaving gases is 2.52 mol.

Amount of steam added in the evaporator $= \dfrac{2.52 \times 39}{61} = 1.61\ \text{mol}$

Weight of steam added $= 1.61 \times 18 = 28.98\ \text{g}$

Amount of outgoing gases = number of moles of incoming gases + number of moles of steam added

$$= 2.52 + 1.61 = 4.13\ \text{mol}$$

Volume of outgoing gases at 358 K and 740 mm Hg $= \dfrac{nRT}{P}$

$$= \frac{4.13 \times 0.082\,05 \times 358 \times 760}{740} = 124.59\ \text{L}$$

3F An evaporator is fed continuously with 25 tonnes per hour of a solution that contains 10% sodium chloride,10% sodium hydroxide and 80% water. During the evaporation process, water is boiled away from the solution, and sodium chloride precipitates out as crystals, which are settled and removed by filtration from the remaining thick liquor. The thick liquor leaving the evaporator contains 50% sodium hydroxide, 1% sodium chloride and 49% water. Calculate (i) the amount of water evaporated in kilograms per hour, (ii) the amount of salt produced in kilograms per hour, and (iii) the amount of thick liquor in kilograms per hour.

Solution

Basis: The process is a flow process. Operation for a particular period of time is the apt basis. For this problem one hour operation is the best basis. The operation involves a liquid system and should be solved on weight basis. The valid assumption for the composition is weight per cent.

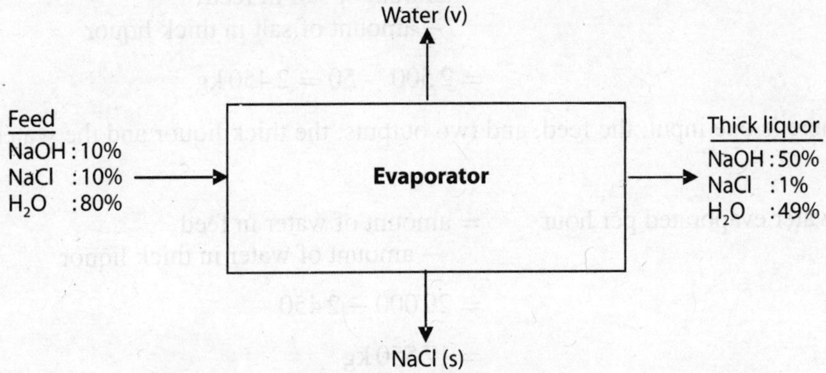

Fig. 3.6 Evaporator

Figure 3.6 represents the data provided in the form of a block diagram.

Weight of solution fed to the evaporator $= 25\,000$ kg

It contains 10% sodium chloride, 10% sodium hydroxide and 80% water.

Weight of sodium hydroxide in the feed $= \dfrac{25\,000 \times 10}{100} = 2\,500$ kg

Weight of sodium chloride in the feed $= \dfrac{25\,000 \times 10}{100} = 2\,500$ kg

Weight of water in the feed $= \dfrac{25\,000 \times 80}{100} = 20\,000$ kg

Sodium hydroxide has one input, that is feed, and only one output, that is thick liquor. All the sodium hydroxide entering the evaporator as a part of feed must leave as a part of the thick liquor.

Weight of sodium hydroxide in the thick liquor $= 2\,500$ kg

Sodium hydroxide is 50% by weight of thick liquor.

Weight of thick liquor produced per hour $= \dfrac{2\,500 \times 100}{50} = 5\,000\,\text{kg}$

Thick liquor contains 1% sodium chloride and 49% water.

Weight of sodium chloride in thick liquor $= \dfrac{5\,000 \times 1}{100} = 50\,\text{kg}$

Weight of water in thick liquor $= \dfrac{5\,000 \times 49}{100} = 2\,450\,\text{kg}$

For sodium chloride, there is one input, the feed, and two outputs: the thick liquor and the salt crystallised out.

Amount of crystalline salt produced per hour = amount of salt in feed − amount of salt in thick liquor

$= 2\,500 - 50 = 2\,450\,\text{kg}$

For water, there is one input, the feed, and two outputs: the thick liquor and the water evaporated as steam.

Amount of water evaporated per hour = amount of water in feed − amount of water in thick liquor

$= 20\,000 - 2\,450$

$= 17\,550\,\text{kg}$

3G A lacquer plant requires 1000 kg of 8% nitrocellulose solution. 5.5% nitrocellulose solution is available. How much of dry nitrocellulose must be added to how much of 5.5% nitrocellulose solution in order to meet the requirement?

Solution

Basis: 1000 kg of 8% nitrocellulose solution

Weight of nitrocellulose in the required solution $= \dfrac{1000 \times 8}{100} = 80\,\text{kg}$

Weight of solvent in the required solution $= 1000 - 80 = 920\,\text{kg}$

The required solution is prepared by mixing together 5.5% nitrocellulose solution and dry nitrocellulose. So the entire solvent for the required solution has to be supplied by the 5.5% nitrocellulose solution.

5.5% nitrocellulose solution contains 5.5% nitrocellulose and 94.5% solvent.

Weight of solvent required for 1000 kg of 8% nitrocellulose solution $= 920\,\text{kg}$

Weight of 5.5% solution required to supply 920 kg of solvent $= \dfrac{920 \times 100}{94.5} = 973.55$ kg

Weight of dry nitrocellulose required $= 1000 - 973.55$

$= 26.45$ kg

3H Two persons calculated the average molecular weight of a flue gas sample containing carbon dioxide, oxygen and nitrogen. The first person used the molecular weight for oxygen and found it to be 30.24; the other person used the atomic weight of oxygen and calculated the average molecular weight to be 28.96. Find the composition of the flue gas.

Solution

Basis: 1 mole of flue gas, as it is a problem involving average molecular weight.

Let the mole fractions of carbon dioxide, oxygen and nitrogen be X_{CO_2}, X_{O_2} *and* X_{N_2}. We have three unknowns and to get them we must have three equations. The equations can be framed based on the following concepts.

- Sum of the mole fractions is equal to one;
- Sum of the multiples of mole fraction of the constituents and their molecular weight is equal to average molecular weight.

$$X_{CO_2} + X_{O_2} + X_{N_2} = 1 \quad (1)$$

$$44X_{CO_2} + 32X_{O_2} + 28X_{N_2} = 30.24 \quad (2)$$

$$44X_{CO_2} + 16X_{O_2} + 28X_{N_2} = 28.96 \quad (3)$$

From equations, (1), (2) and (3) we get

$X_{N_2} = 0.8$

$X_{O_2} = 0.08$

$X_{CO_2} = 0.12$

Table 3.2 Composition of flue gas

Constituent	Volume per cent Mole per cent
CO_2	12
O_2	8
N_2	80

The mole per cent of a gas can be obtained by multiplying the mole fraction by 100. For a gaseous system the composition should be expressed in mole per cent which is also equal to volume per cent. The composition of the flue gas is given in Table 3.2.

3I A batch of soyabeans contains 18% oil and 10% water. It is treated with pure hexane to give a solution containing 25% oil and 75% hexane and a residue containing 0.5% oil and 20% hexane. What weight of soyabeans is required to give 1000 kg of solution?

Solution

Basis: 100 kg of soyabeans. The valid assumption for solids is weight per cent. As 100 kg soyabeans has been chosen as the basis, the weight per cent of each component gives the actual weight of that component in 100 kg of the material.

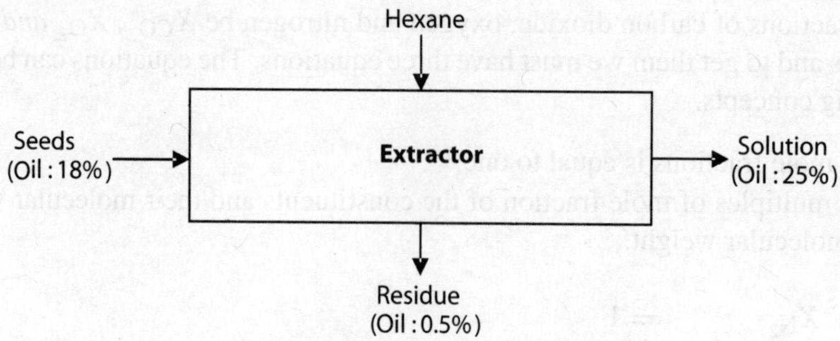

Fig. 3.7 Extractor

Figure 3.7 represents the data provided in the form of a block diagram.

Soyabeans contain 18% oil and 10% water.

Weight of oil in 100 kg soyabeans $= 18$ kg

Weight of inerts and water in 100 kg soyabeans $= 100 - 18 = 82$ kg

For inerts and water there is one input and only one output, the residue. The residue contains 20% hexane, 0.5% oil and the rest consists of inerts and water.

Weight per cent of inerts and water in residue $= 100 - 20 - 0.5$

$= 79.5$wt%

Weight of inerts and water in residue $= 82$ kg

This forms 79.5% of the residue.

Weight of residue $= \dfrac{82 \times 100}{79.5} = 103.14\,\text{kg}$

The residue contains 0.5% oil. This oil cannot be recovered.

Weight of oil in residue $= \dfrac{103.14 \times 0.5}{100} = 0.52\,\text{kg}$

Weight of oil recovered from 100 kg soyabeans $= 18 - 0.52 = 17.48\,\text{kg}$

Oil is recovered as a solution containing 25% oil and 75% hexane.

Weight of solution obtained from 100 kg soyabeans $= \dfrac{17.48 \times 100}{25} = 69.92\,\text{kg}$

For 69.92 kg of solution, weight of soyabeans required = 100 kg

For 1000 kg of solution, weight of soyabeans required $= \dfrac{1000 \times 100}{69.92} = 1\,430.3\,\text{kg}$

3J 100 kg of a mixture of $Na_2SO_4 \cdot 10H_2O$ and $Na_2CO_3 \cdot 10H_2O$ is heated to drive off the water of hydration. The final weight of the mixed salts is 40.58 kg. What is the molar ratio of sodium carbonate to sodium sulphate in the original hydrated salts.

Solution

Basis: 100 kg of the mixture

Let the weight. of $Na_2SO_4 \cdot 10H_2O$ be X. Then the weight of $Na_2CO_3 \cdot 10H_2O$ will be $100 - X$.

We have introduced one unknown. To get its value, one equation is required. The required equation can be framed as explained below.

Molecular weight of $Na_2SO_4 \cdot 10H_2O$ is 322 and that of Na_2SO_4 is 142.

322 g of hydrated Na_2SO_4 on dehydration by heating give 142 g of Na_2SO_4.

Molecular weight of $Na_2CO_3 \cdot 10H_2O$ is 286 and that of Na_2CO_3 is 106.

286 g of hydrated Na_2CO_3 on dehydration by heating give 106 g of Na_2CO_3.

Total weight of anhydrous salt mixture
obtained by heating 100 kg of hydrated salt mixture $= \dfrac{142 \times X}{322} + \dfrac{106 \times (100 - X)}{286}$

This is given in the problem as 40.58. So the required equation is

$$\frac{142 \times X}{322} + \frac{106 \times (100 - X)}{286} = 40.58$$

Solving, the weight of $Na_2SO_4 \cdot 10H_2O$ $= 49.71\,kg = \dfrac{49.71}{322} = 0.154\,kmol$

Weight of $Na_2CO_3 \cdot 10H_2O$ $= 50.29\,kg = \dfrac{50.29}{286} = 0.176\,kmol$

Molar ratio of $Na_2CO_3 \cdot 10H_2O$ to $Na_2SO_4 \cdot 10H_2O$ $= 0.176 : 0.154$

$= 1.143 : 1.0$

3K A solution contains 25% salt in water. (i) Express the composition in kilograms of water per kilogram of salt and kilograms of salt per kilogram of water. (ii) What is the composition of the solution in weight per cent of salt if 40% of water originally present is evaporated? (iii) What is the composition if an amount of water equal to 40% of the weight of the original solution is evaporated? (iv) How much of water must be evaporated from 100 kg of solution if the final composition has to be 40wt% water?

Solution

Basis: 100 kg of solution. The valid assumption for liquid systems is weight per cent. As 100 kg of solution has been chosen as the basis, the weight per cent of each component gives the actual weight of that component.

Weight of salt in 100 kg of solution $= 25\,kg$

Weight of water in 100 kg of solution $= 75\,kg$

(i) Composition in kilograms of water per kilogram of salt $= \dfrac{75}{25} = 3.0\,kg$

Composition in kilograms of salt per kilogram of water $= \dfrac{25}{75} = 0.333\,kg$

(ii) Weight of water present originally $= 75\,kg$

Weight of water evaporated (40% of water is evaporated) $= \dfrac{75 \times 40}{100} = 30\,kg$

Weight of remaining water $= 75 - 30 = 45\,kg$

Weight of final solution $=$ weight of salt $+$ weight of water

$= 25 + 45 = 70\,kg$

Composition of solution in weight per cent of salt $= \dfrac{25 \times 100}{70} = 35.71\,wt\%$

(iii) Weight of original solution $= 100\,kg$

Weight of water evaporated, if 40% solution is to be evaporated $= \dfrac{100 \times 40}{100} = 40\,\text{kg}$

Weight of remaining water $= 75 - 40 = 35\,\text{kg}$

Weight of final solution = weight of salt + weight of water

$= 25 + 35 = 60\,\text{kg}$

Composition of solution in weight per cent of salt $= \dfrac{25 \times 100}{60} = 41.67\,\text{wt\%}$

(iv) Weight of solution = 100 kg (25 kg salt and 75 kg water)

Water is evaporated till the weight per cent of water in the resultant solution becomes 40%. The remaining 60% is salt, which weighs 25 kg.

Weight of water in the resultant solution $= \dfrac{25 \times 40}{60} = 16.67\,\text{kg}$

Weight of water to be evaporated = weight of water originally present − weight of water in the final solution

$= 75 - 16.67 = 58.33\,\text{kg}$

3L A gas mixture containing 28% carbon dioxide and 72% ammonia is scrubbed with an acid solution to remove ammonia. The resulting gas mixture contains 30% ammonia. What percentage of the original ammonia has been removed? It may be assumed that carbon dioxide is not absorbed by the acid solution and there is no evaporation in the scrubber.

Solution

Basis: 100 mol of gas mixture. The valid assumption for a gaseous system is mole per cent. As 100 mol of gaseous mixture has been chosen as the basis, the mole per cent of each component gives the actual number of moles of that component.

Amount of carbon dioxide in 100 mol of gas mixture = 28 mol

Amount of ammonia in 100 mol of gas mixture = 72 mol

When the gas mixture is scrubbed with the acid solution, carbon dioxide is not affected. So the amount of carbon dioxide in the outlet gas mixture will also be 28 mol. It is given that the outlet gas from the scrubber contains 30% ammonia and the balance of 70% is carbon dioxide. Hence 70% of the outlet gas is 28 mol.

Amount of ammonia in the outlet gas $= \dfrac{28 \times 30}{70} = 12\,\text{mol}$

Amount of ammonia absorbed = amount of ammonia in the inlet gas − amount of ammonia in the outlet gas

$= 72 - 12 = 60\,\text{mol}$

Percentage of ammonia absorbed $= \dfrac{60 \times 100}{72} = 83.33\%$

3M A distillation column separates a mixture of ethanol and water into high purity alcohol and wastewater. The feed is a 20mol% ethanol and 80mol% water mixture. The distillate is to contain 85mol% ethanol and the residue is to have 3mol% ethanol. Compute (i) the quantity of distillate and residue produced per 1000 kg of feed, and (ii) the percentage of alcohol in the feed that has been recovered as distillate.

Solution

Basis: 1000 kg of feed

Compositions of all the streams have been given in mole per cent. So the amount of feed has to be known on molar basis. If the weight in kilograms is divided by the average molecular weight of the feed, the amount of feed in kilomoles will be known. The sum of the multiples of the mole fraction of each constituent and its molecular weight is equal to the average molecular weight. The mole fraction of a component is obtained by dividing its composition in mole per cent by 100.

The molecular weight of ethanol is 46 and its mole fraction in the feed mixture is 0.2.

The molecular weight of water is 18 and its mole fraction in the feed mixture is 0.8.

The average molecular weight of the feed $= 46 \times 0.2 + 18 \times 0.8 = 23.6$

Amount of feed $= \dfrac{1000}{23.6} = 42.37\,\text{mol}$

The distillation column splits the feed (F) into distillate (D) and residue (R). We have two unknowns and two equations are required to get the value of the unknowns.

Feed = Distillate + Residue

$F = D + R$

Feed, distillate and residue are all mixtures of ethanol and water. Ethanol balance, which equates the amount of ethanol in the feed with that in the distillate and the residue can be used as a second equation.

$$FX_F = DX_D + RX_R$$

X_F is the mole fraction of ethanol in the feed, X_R is the mole fraction of ethanol in the residue, and X_D is the mole fraction of ethanol in the distillate.

$$D + R = 42.37 \quad (1)$$

$$0.85D + 0.03R = 42.37 \times 0.2 = 8.67 \quad (2)$$

From equations (1) and (2),

$$R = 33.34 \text{ kmol}$$

$$D = 9.03 \text{ kmol}$$

The feed contains 20% ethanol and the distillate contains 85% ethanol.

Amount of ethanol in the feed $= 42.37 \times \dfrac{20}{100} = 8.47$ kmol

Amount of ethanol in the distillate $= 9.03 \times \dfrac{85}{100} = 7.68$ kmol

Out of 8.47 kmol of ethanol in the feed, 7.68 kmol of ethanol has been recovered as a component of the distillate.

Percentage of ethanol recovered $= \dfrac{7.68 \times 100}{8.47} = 90.67\%$

3N It is desired to produce a mixed acid containing 40% sulphuric acid, 35% nitric acid and 25% water from waste acid containing 37% sulphuric acid, 23% nitric acid and 40% water, using 80% concentrated sulphuric acid and 70% concentrated nitric acid. Calculate the weight of waste acid, concentrated sulphuric acid and concentrated nitric acid required to make 4 000 kg of the desired acid.

Solution

Basis: 4 000 kg of the desired acid mixture. It is produced by mixing together the waste acid, concentrated sulphuric acid and concentrated nitric acid. All the three input streams as well as the final output are liquid streams. So the valid assumption for expressing their composition is weight per cent. Dividing the weight per cent by 100 gives the weight fraction.

Figure 3.8 represents the data provided in the form of a block diagram.

Let X be the weight of waste acid, Y the weight of concentrated sulphuric acid and Z the weight of concentrated nitric acid required to produce 4 000 kg of the desired acid. We have introduced three unknowns. Three equations are therefore required to get the value of these unknowns. The required equations can be framed as explained below.

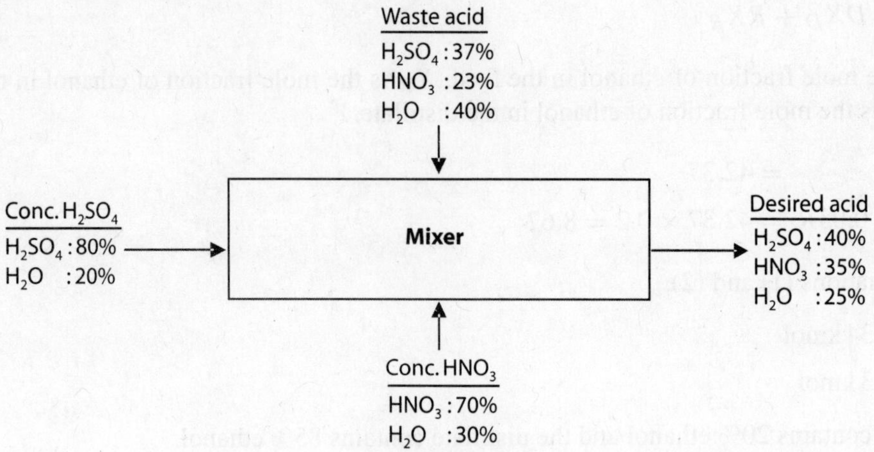

Fig. 3.8 Mixer

Overall balance Equating the weights of the three inputs with the weight of the output we get,

$$X + Y + Z = 4\,000 \tag{1}$$

Sulphuric acid balance Equating the weight of sulphuric acid in the input with the weight of sulphuric acid in the output, we can get the second equation. Sulphuric acid is present only in the waste acid and concentrated sulphuric acid.

$$0.37X + 0.8Y = 4\,000 \times 0.4 = 1\,600 \tag{2}$$

Nitric acid balance Equating the weight of nitric acid in the input with the weight of nitric acid in the output, we can get the third equation. Nitric acid is present only in the waste acid and concentrated nitric acid.

$$0.23X + 0.7Z = 4\,000 \times 0.35 = 1\,400 \tag{3}$$

From equations (1), (2) and (3),

Weight of waste acid required (X) $= 0$ kg

Weight of concentrated sulphuric acid (Y) $= 2\,000$ kg

Weight of concentrated nitric acid required (Z) $= 2\,000$ kg

30 A saturated solution of a salt is made by agitating an excess of salt in water and draining off the solution. When this is done a certain quantity of solution remains adsorbed on the undissolved salt. If 1 kg of water will dissolve 0.25 kg of salt, and if 0.5 kg of solution will remain adsorbed on 1 kg of undissolved salt, calculate (i) the amount of solution that can be drained off when 1000 kg of water is mixed with 500 kg of salt, and (ii) the amount of dry salt obtained if the undissolved wet salt is thoroughly dried.

Solution

Basis: 1000 kg of water and 500 kg of salt are mixed

1 kg of water dissolves 0.25 kg of salt.

Weight of salt dissolved in 1000 kg water $= 1000 \times 0.25 = 250$ kg

Weight of solution formed = weight of water + weight of dissolved salt

$= 1\,250$ kg

Weight of undissolved salt = weight of salt taken − weight of salt dissolved

$= 500 - 250 = 250$ kg

1 kg of undissolved salt absorbs 0.5 kg of solution.

Weight of solution adsorbed by undissolved salt $= 250 \times 0.5 = 125$ kg

Weight of solution that can be drained off = weight of solution formed − weight of solution absorbed by undissolved salt

$= 1\,250 - 125 = 1\,125$ kg

1 250 kg solution contains 250 kg salt.

Weight of salt in the solution absorbed by undissolved salt $= \dfrac{125 \times 250}{1\,250} = 25$ kg

Weight of dry salt obtained when wet undissolved salt is dried = weight of undissolved salt + weight of salt in the solution absorbed by undissolved salt

$= 250 + 25 = 275$ kg

3P A caustic soda unit has a contract to buy 98% pure sodium chloride for Rs. 2 000 per tonne delivered. A shipment of 10 tonnes was only 90% pure. Calculate (i) the amount to be paid for this shipment, and (ii) how much of this 90% pure material should be added to 2 m^3 water to prepare 5wt% sodium chloride solution.

Solution

(i) *Basis*: 10 tonnes of 90% pure sodium chloride delivered

Weight of sodium chloride delivered $= 10\,000 \times 0.9 = 9\,000$ kg

Cost of 1 tonne of 98% sodium chloride or 980 kg sodium chloride $= \text{Rs } 2\,000$

Cost of 9 000 kg sodium chloride $= \dfrac{2\,000 \times 9\,000}{980} = \text{Rs. } 18\,367$

(ii) *Basis*: 2 m^3 water weighing 2 000 kg, as the density of water is 1000 kg m^{-3}

5 wt% NaCl solution is required. Water in the solution is 95% by weight, which is 2 000 kg.

Weight of 100% pure NaCl required $= \dfrac{2\,000 \times 5}{95} = 105.26 \text{ kg}$

Weight of 90% pure NaCl $= \dfrac{105.26 \times 100}{90} = 116.96 \text{ kg}$

3Q A sample of coal contains 79% by weight of carbon and 6% by weight of ash. The clinker from the furnace refuse pit contains 90% ash and 10% carbon. Based on 100 kg coal charged, find the weight in kilograms of refuse produced, and percentage of fuel value lost.

Solution

Basis: 100 kg coal charged

Weight of carbon in 100 kg coal $= 79 \text{ kg}$

Weight of ash in 100 kg coal $= 6 \text{ kg}$

The furnace refuse contains 90% ash and 10% carbon.

Weight of ash in refuse produced $= 6 \text{ kg}$

Ash forms 90% of refuse.

Weight of refuse produced $= \dfrac{6 \times 100}{90} = 6.67 \text{ kg}$

Weight of carbon in refuse = weight of refuse – weight of ash

$= 6.67 - 6.00 = 0.67 \text{ kg}$

Out of 79 kg carbon, 0.67 kg is lost.

Percentage of fuel value lost $= \dfrac{0.67 \times 100}{79} = 0.85\%$

3R A certain soap contains 60% moisture on wet basis when raw. Before it can be pressed into cakes for sale, enough water must be evaporated to reduce the moisture content to 20%, also on wet basis. How many 125 g cakes of soap can be pressed from one tonne of original raw soap? Neglect cutting waste.

Solution

Basis: 1 tonne, that is 1000 kg raw soap

Raw soap contains 60% moisture and the balance of 40% is dry soap.

Weight of dry soap in 1000 kg raw soap $= \frac{1000 \times 40}{100} = 400$ kg

After evaporation of a portion of water, the finished soap ready for pressing contains 20% moisture. It means that 80% is dry soap, the weight of which is 400 kg.

Weight of finished soap $= \frac{400 \times 100}{80} = 500$ kg

Number of 125 g soap cakes that can be pressed out $= \frac{500 \times 1000}{125} = 4000$

3S A company produces a dairy product containing 25% moisture. They sell this at the rate of Rs 80 per kilogram. The company now comes up with an improved product containing 10% moisture. If the drying cost is Rs 80 per kilogram of water removed, and the same profit margin has to be maintained, what should be the selling cost?

Solution

Basis: 1 kg of the original product

Original product contains 25% moisture.

Weight of moisture-free substance in 1 kg of original product $= 0.75$ kg

The improved product contains 10% moisture and moisture-free substance forms 90%.

90% weight of improved product is 0.75 kg.

Weight of improved product that can be obtained from 1 kg of original product $= \frac{0.75 \times 100}{90}$

$= 0.833$ kg

Weight of water evaporated for producing an improved product $= 1.0 - 0.833$

$= 0.167$ kg

Cost of water evaporation is Rs 80 per kilogram of water.

Cost of evaporating water from the original product $= 0.167 \times 80 =$ Rs 13.36

Total cost $=$ Rs 80 $+$ Rs 13.36 $=$ Rs 93.36

1 kg of original product yields only 0.833 kg of improved product.

Selling cost per kg for the improved product,

if the same profit margin has to be maintained $= \frac{93.36}{0.833} =$ Rs 112.08

Problems for Practice

3.1 Wet material containing 10% moisture is to be dried to contain 2% moisture by circulation of hot air. The inlet humidity is 0.02 g and the exit humidity is 0.09 g water per gram of dry air. Find the volume of fresh air required if 2 000 kg of the product is desired per hour. Air enters at 303 K and atmospheric pressure.

3.2 30 m^3 of moist air at a total pressure of 101.325 kPa and 303 K contains water vapour in such a proportion that its partial pressure is 2.933 kPa. Keeping the total pressure constant, the temperature is reduced to 288 K, during which process, some water vapour condenses out. After cooling, the partial pressure of water vapour is found to be 1.693 kPa. Calculate the volume of air at 288 K and the weight of water condensed.

3.3 Combustion gases having the following molal composition are passed into an evaporator at 573 K and 743 mm Hg (nitrogen 79.2%, oxygen 7.2% and carbon dioxide 13.6%). Water is added to the stream as vapour and the gases leave at 373 K and 740 mm Hg with the following composition: nitrogen 48.3%, oxygen 4.4% and water vapour 39%. Calculate (i) the volume of gases leaving the evaporator per 100 L of gases entering, and (ii) the weight of steam added to the gases.

3.4 A mixture is separated into two fractions. The first fraction contains 50% of A and 50% of B. The second fraction contains 25% of A and 75% of B. The weight of the second fraction is twice that of the first. What is the composition of the original mixture?

3.5 An aqueous solution containing 10% sodium chloride and 10% potassium chloride is desired. A solution is made up but it was found to contain only 5% sodium chloride and 4% potassium chloride. To bring the concentrations to the desired value, one concentrated solution containing 30% potassium chloride and another containing 25% sodium chloride were added. How much of each solution must be used to produce 1000 kg of the desired solution? All percentages are by weight.

3.6 Waste acid containing 25% nitric acid, 50% sulphuric acid and 25% water by weight is to be concentrated to contain 30% nitric acid and 60% sulphuric acid by adding concentrated sulphuric acid (95% by weight) and concentrated nitric acid (90% by weight). Calculate the weights of waste acid and concentrated acids to obtain 10 000 kg of the mixture of desired composition.

3.7 A mixture containing 30% nitric acid, 40% sulphuric acid and 30% water has to be made by mixing 98% pure aqueous sulphuric acid, 90% pure aqueous nitric acid and a waste acid containing 5% nitric acid, 20% sulphuric acid and the balance water. Calculate the weight of each of the above three streams required for making 1000 kg of the required mixture.

3.8 A solution of iodine in glycerin is prepared by dissolving iodine crystals in glycerin. The solubility of iodine in glycerin is 1.95% by weight in 95% glycerin (95 parts of glycerin and 5 parts water). Glycerin of 99% concentration is available. Water is added to this to obtain a 95% solution. Some iodine is lost while handling due to sublimation. Laboratory tests reveal that average loss of iodine is 0.02% by weight. Compute the quantities of the various ingredients required to make 500 kg of iodine solution.

3.9 The rate at which waste gases are leaving a stack is to be determined by adding a known quantity of carbon dioxide to the gases at a point between the fire box and base of the stack. Waste gases from the fire box contain 3.5% carbon dioxide. Pure carbon dioxide is added at the rate of 2.2 kg per minute. The gases leaving the top of the stack contain 4.2% carbon dioxide. Calculate the rate of flow of gases from the fire box in kmol h^{-1}.

3.10 In the separation of a gaseous mixture by diffusion through a plastic barrier the following results were obtained.

- Composition of the mixture fed to the barrier: 50% oxygen and 50% helium.
- Composition of the mixture diffused through the barrier: 40% oxygen and 60% helium.
- 30% of the gas fed to the barrier diffused through it.

Calculate the composition of the gaseous mixture that did not diffuse through the barrier.

3.11 10 000 kg of a 20% solution of sodium nitrate is concentrated to saturation. Crystals of sodium nitrate are removed by filtration. 0.05 kg of solution adheres to each kilogram of crystals. When the crystals are dried, sodium nitrate in the adhering solution is deposited on the crystals. The solubility of sodium nitrate in water is 1.76 kg per kilogram of water at 373 K and 0.88 kg per kilogram of water at 293 K. Calculate (i) the amount of water to be evaporated, and (ii) the weight of crystals obtained.

3.12 10 kmol of carbon dioxide are added every hour to a stream of pure air. The resultant mixture contains 7.5mol% carbon dioxide. How many kilomoles of gas mixture are flowing per hour? How many moles per minute of oxygen are flowing past a point? Ignore the carbon dioxide content of air.

3.13 Phosphate rock costing Rs 5 000 per tonne is added to a cheap fertilizer to enrich it. If the cheap fertiliser costs Rs 2 000 per tonne and the final product (including 30% blender's profit) is to be marketed at Rs 4 000 per tonne, how much of phosphate rock should be added to each tonne of cheap fertilizer?

4 Material Balance in Reaction Systems

Almost all manufacturing processes in the chemical industry involve chemical reactions. In industrial reactions, exact stoichiometric amounts of reactants are rarely used. To make a desired reaction take place or to use a costly reactant completely, excess reactants are nearly always used.

Limiting reactant is the reactant which is present in smallest stoichiometric amount.

Excess reactant is the reactant which is present in excess over and above the stoichiometrically required amount to completely react with the limiting reactant.

Percentage excess is the percentage ratio of the excess number of moles of excess reactant to the theoretical number of moles of excess reactant required to react completely with the limiting reactant.

The excess reactant comes out together with the product. Sometimes it is separated out and used again. If the reaction is not complete or there are side reactions, the products will be accompanied by unused reactants as well as by-products. In this context the following definitions assume significance.

Conversion is the percentage of feed, or some material in the feed, that is converted to products.

Degree of completion of reaction is the percentage of limiting reactant converted to desired product.

Yield is the percentage ratio of the number of moles of the limiting reactant undergoing a desired reaction to the number of moles of limiting reactant reacted.

Selectivity is the ratio of the number of moles of the reactant undergoing a desired reaction to the number of moles of reactant undergoing an undesired reaction.

The following example explains the above definitions more precisely.

Example

Consider the chlorination of benzene to produce monochlorobenzene. Let us assume that in an experimental run, 2 mol of benzene and 3 mol of chlorine react to give 1.8 mol of monochlorobenzene and 0.1 mol of dichlorobenzene.

The reactions occurring are

$$C_6H_6 + Cl_2 \rightarrow C_6H_5Cl + HCl \quad (1)$$

$$C_6H_6 + 2Cl_2 \rightarrow C_6H_4Cl_2 + 2HCl \quad (2)$$

Reaction (1) is the desired reaction and reaction (2) is the undesired side reaction. For reaction (1), 1 mol of benzene requires 1 mol of chlorine. The feed here is 2 mol of benzene and 3 mol of chlorine. Benzene is present in lower stoichiometric amount. So it is the **limiting reactant** and chlorine is the **excess reactant**. 2 mol of benzene require 2 mol of chlorine and 1 mol of chlorine is in excess.

$$\text{Percentage excess of chlorine} = \frac{1}{2} \times 100 = 50\%$$

Degree of completion is the percentage ratio of the number of moles of the limiting reactant that have undergone the desired reaction to the number of moles of the limiting reactant in the feed. Of the 2 mol of benzene fed, 1.8 mol of benzene have undergone the desired reaction to produce 1.8 mol of monochlorobenzene.

$$\text{Degree of completion of reaction} = \frac{1.8}{2} \times 100 = 90\%$$

Of the 2 mol of benzene fed, 0.1 mol of benzene has undergone the undesired side reaction to produce 0.1 mol of dichlorobenzene. In all 1.9 mol of benzene have reacted.

$$\text{Conversion of benzene} = \frac{1.9}{2.0} \times 100 = 95\%$$

1 mol of chlorine is required for the formation of 1 mol of monochlorobenzene, and 2 mol of chlorine are required for the formation of 1 mol of dichlorobenzene.

$$\text{Number of moles of chlorine reacted} = (1.8 \times 1) + (0.1 \times 2) = 2.0\text{ mol}$$

$$\text{Conversion of chlorine} = \frac{2.0}{3.0} \times 100 = 66.67\%$$

It may be noted that conversion based on excess reactant will be less than the degree of completion of reaction.

1.8 mol of monochlorobenzene is obtained for 1.9 mol of benzene reacted.

$$\text{Yield} = \frac{1.8}{1.9} \times 100 = 94.74\%$$

Of the 1.9 mol of benzene reacted, 1.8 mol have undergone the desired reaction and 0.1 mol has undergone the undesired reaction. Selectivity is the ratio of the number of moles of the reactant undergoing a desired reaction to the number of moles of reactant undergoing an undesired reaction.

$$\text{Selectivity} = \frac{1.8}{0.1} = 18$$

When a chemical reaction takes place, the individual chemical compounds change in amount. It is no longer possible to make a balance for a particular chemical compound. The two quantities that are conserved during a chemical transformation are total mass and elemental atoms. An elemental balance becomes the key to solving this type of problem. This balance is easily obtainable from a complete knowledge of chemical reactions involved.

Note: The guidelines for solving problems given in the previous chapter hold good here also.

ILLUSTRATIONS

4A 300 kg of nitrogen and 75 kg of hydrogen are brought together and allowed to react at 823 K and 300 atm. It is found that there are 38 kmol of gases present at equilibrium. How many kilomoles of nitrogen, hydrogen and ammonia are present at equilibrium? Which is the limiting reactant? What is the percentage excess? What is the degree of completion? What is the percentage conversion of hydrogen to ammonia?

Solution

Basis: 300 kg of nitrogen and 75 kg of hydrogen fed to the reactor

$$\text{Amount of nitrogen fed to the reactor} = \frac{300}{28} = 10.71\,\text{kmol}$$

$$\text{Amount of hydrogen fed to the reactor} = \frac{75}{2.016} = 37.20\,\text{kmol}$$

The reaction occurring in the reactor is

$$N_2 + 3H_2 \rightleftharpoons 2NH_3$$

1 mol of nitrogen requires 3 mol of hydrogen.

Theoretically hydrogen required to consume all the nitrogen $= 10.71 \times 3 = 32.13$ kmol

As nitrogen is present in lower stoichiometric amount, it is the limiting reactant. Hydrogen is the excess reactant.

Excess of hydrogen = hydrogen fed − theroretically required hydrogen
$= 37.20 - 32.13 = 5.07$ kmol

$$\text{Percentage of excess hydrogen} = \frac{5.07 \times 100}{32.13} = 15.80\%$$

At equilibrium, let X mol of nitrogen react with $3X$ mol of hydrogen to give $2X$ mol of ammonia.

Amount of unreacted hydrogen $= (37.20 - 3X)$ kmol

Amount of unreacted nitrogen $= (10.71 - X)$ kmol

Amount of ammonia formed $= 2X$ kmol

Total amount of gas at equilibrium $= (47.91 - 2X)$ kmol

Total amount of gas at equilibrium is given in the problem as 38 kmol.

$\therefore (47.91 - 2X) = 38$

$X = 4.955$

At equilibrium

Amount of hydrogen present = 22.335 kmol

Amount of nitrogen present = 5.755 kmol

Amount of ammonia present = 9.91 kmol

Degree of completion of reaction is based on the limiting reactant, nitrogen. Out of 10.71 kmol nitrogen fed, 4.955 kmol have reacted.

$$\text{Degree of completion of reaction} = \frac{4.955 \times 100}{10.71} = 46.27\%$$

Out of 37.20 kmol hydrogen fed, 14.865 kmol reacted.

$$\text{Percentage conversion of hydrogen} = \frac{14.865 \times 100}{37.20} = 39.96\%$$

4B A lamp burns using acetylene produced according to the equation

$$CaC_2 + 2H_2O \rightarrow C_2H_2 + Ca(OH)_2$$

If the lamp consumes 100 litres of gas per hour at 298 K and a pressure of 750 Torr, compute the weight of calcium carbide required for 15 hours service of the lamp.

Solution

Basis: 15 hours service life of lamp

Acetylene gas required per hour = 100 L at 298 K, 750 Torr

Acetylene gas required for 15 hours = 1500 L at 298 K, 750 Torr

$$\text{Amount of acetylene gas required is given by } n = \frac{PV}{RT}$$

$$= \frac{750 \times 1500}{760 \times 0.08205 \times 298} = 60.54 \text{ mol}$$

As per the reaction given, 1 mol of calcium carbide gives 1 mol of acetylene.

Calcium carbide required $= 60.54 \text{ mol} = (60.54 \times 64) \text{ g}$

$= 3875 \text{ g} = 3.875 \text{ kg}$

(Molecular weight of calcium carbide is 64.)

4C Acetylene is manufactured using limestone. Limestone on analysis is found to contain 95% calcium carbonate and the rest inerts. This material is burnt in a limekiln to produce calcium oxide and the decomposition reaction goes to completion. The solid products are reacted in an electric furnace with dry powdered coke of 90% carbon. The reaction taking place is

$$CaO + 3\,C \rightarrow CaC_2 + CO$$

and it goes to 85% completion, if 15% excess calcium oxide is used. Solid products from the furnace are treated with water to get acetylene and calcium oxide. This reaction goes to completion. Calculate the weight of limestone and coke required for the production of ten tonnes of acetylene per day.

Solution

Basis: Production of 10 tonnes of acetylene per day

The reactions occurring are given below. Purity of the various components and the percentage excess of the reactants, the degree of completion and the molecular weight of the components are also indicated.

$$\text{(i)}\quad \underset{100}{\overset{95\%\text{ pure}}{CaCO_3}} \xrightarrow[\text{completion}]{100\%} \underset{56}{CaO} + \underset{44}{CO_2}$$

$$\text{(ii)}\quad \underset{56}{\overset{15\%\text{ excess}}{CaO}} + \underset{36}{\overset{90\%\text{ pure}}{3\,C}} \xrightarrow[\text{completion}]{85\%} \underset{64}{CaC_2} + \underset{28}{CO}$$

$$\text{(iii)}\quad \underset{64}{CaC_2} + \underset{18}{H_2O} \xrightarrow[\text{completion}]{100\%} \underset{26}{C_2H_2} + \underset{56}{CaO}$$

Calcium carbide

26 g of acetylene is obtained from 64 g of calcium carbide.

$$\text{Calcium carbide required for 10 000 kg of acetylene} = \frac{10\,000 \times 64}{26} = 24\,615.4\ \text{kg}$$

Calcium carbide is produced in reaction (ii) when 15% excess calcium oxide is used and the reaction goes to 85% completion.

Calcium oxide

56 g of calcium oxide gives 64 g of calcium carbide.

$$\text{Calcium oxide required} = \frac{24\,615.4 \times 56}{64} \times \frac{1.15}{0.85} = 29\,140.3\ \text{kg}$$

Calcium oxide is produced by burning limestone. 56 g of calcium oxide is obtained from 100 g of limestone and limestone is 95% pure.

$$\text{95\% pure limestone required} = \frac{29\,140.3 \times 100}{56 \times 0.95} = 54\,775\ \text{kg}$$

Coke

Coke is required for the production of calcium carbide as per reaction (ii). Coke is 90% pure and the reaction goes to 85% completion. 64 g of calcium carbide requires 36 g of coke.

90% pure coke required for the production of 24 614.4 kg calcium carbide

$$= \frac{24\,615.4 \times 36}{64} \times \frac{1}{0.85 \times 0.90}$$

$$= 18\,099.6\ \text{kg}$$

4D Nitrobenzene is produced commercially by nitrating benzene with a mixed acid containing 39% nitric acid, 53% sulphuric acid and 8% water. A charge is made up of 655 kg benzene and 1 360 kg of mixed acid. The reaction is 95% complete. Calculate (i) the quantity of nitrobenzene produced, and (ii) the quantity and quality of the spent acid.

Solution

Basis: 655 kg benzene and 1 360 kg of mixed acid charged to the reactor.

The reaction mixture is first split into its components. The amount of reactants in moles is calculated.

$$\text{Amount of benzene fed} = \frac{655}{78} = 8.40\ \text{kmol}$$

Amount of nitric acid in the mixed acid $= \dfrac{1\,360 \times 39}{100} = 530.4\,\text{kg}$

$= \dfrac{530.4}{63} = 8.42\,\text{kmol}$

Amount of sulphuric acid in the mixed acid $= \dfrac{1\,360 \times 53}{100} = 720.8\,\text{kg}$

Amount of water in the mixed acid $= \dfrac{1\,360 \times 8}{100} = 108.8\,\text{kg}$

The reaction mixture contains 8.40 kmol of benzene and 8.42 kmol of nitric acid. The reaction is

$$C_6H_6 + HNO_3 \quad \rightarrow \quad C_6H_5NO_2 + H_2O$$

Benzene is the limiting reactant as it is present in lower stoichiometric amount. The reaction goes to 95% completion. The degree of completion is based on limiting reactant.

Benzene reacted $= 8.4 \times 0.95 = 7.98\,\text{kmol}$

1 mol of benzene reacts with 1 mol of nitric acid to give 1 mol of nitrobenzene and 1 mol of water.

Nitrobenzene formed $= 7.98\,\text{kmol}$

$= 7.98 \times 123 = 981.54\,\text{kg}$

Nitric acid reacted $= 7.98\,\text{kmol}$

Excess of nitric acid = nitric acid fed – nitric acid reacted

$= 8.42 - 7.98 = 0.44\,\text{kmol}$

$= 0.44 \times 63 = 27.72\,\text{kg}$

Water formed $= 7.98\,\text{kmol}$

$= 7.98 \times 18 = 143.64\,\text{kg}$

Water originally present $= 108.80\,\text{kg}$

Total amount of water present $= 252.44\,\text{kg}$

The mixture of unreacted nitric acid, inert sulphuric acid and water formed in the reaction and water originally present is known as spent acid. The composition of the spent acid is given in Table 4.1.

Table 4.1 Composition of the spent acid

Constituent	Weight (kg)	Weight per cent
Unreacted HNO_3	27.72	2.77
H_2O	252.44	25.22
Inert H_2SO_4	720.80	72.01
	1000.96	

4E In a process for the manufacture of acetylene, pure oxygen and pure methane are fed to the burner. The cracked gas from the burner has the following composition: hydrogen 56.5%, methane 5.2%, ethylene 0.3%, acetylene 7.5%, propene 0.5%, carbon monoxide 25.8%, carbon dioxide 4.0%, oxygen 0.2% by mole on dry basis. For 100 kmol of cracked gas, calculate methane requirement, oxygen requirement, production of water, conversion of methane and yield of acetylene.

Solution

Basis: The composition and rate of input are not precisely known. The problem has to be solved based on the output, even though composition of output is partially known. The oxidation product of a gaseous mixture containing hydrogen is bound to have water vapour. Analysis of the output from the reactor is given on dry basis. However, the given data are sufficient to extract the unknown information. So 100 kmol dry cracked gas is chosen as the basis.

Pure methane and oxygen are fed to the burner. The feed is cracked into a mixture of hydrocarbons, carbon oxides, hydrogen and oxygen. As 100 kmol has been chosen as the basis, mole per cent of each component also gives the number of kilomoles of that particular component.

The problem can be easily solved based on an elemental balance for the process. Table 4.2 gives the number of moles of carbon, hydrogen and oxygen atoms present in the dry cracked gas.

Table 4.2 Composition of dry cracked gas

Components	Kilomoles (kmol)	H (kmol)	C (kmol)	O (kmol)
H_2	56.5	113	–	–
CH_4	5.2	20.8	5.2	–
C_2H_4	0.3	1.2	0.6	–
C_2H_2	7.5	15.0	15.0	–
C_3H_6	0.5	3.0	1.5	–
CO	25.8	–	25.8	25.8
CO_2	4.0	–	4	8
O_2	0.2	–	–	0.2
		153.0	52.1	34.2

The source of carbon and hydrogen in the cracked gas is methane and the source of oxygen in it is oxygen.

100 kmol of dry cracked gas contain 52.1 kmol of carbon atoms.

One molecule of methane contains one atom of carbon.

Methane required $= 52.1$ kmol

In 1 molecule of methane, 1 carbon atom is accompanied by 4 hydrogen atoms.

Hydrogen accompanying carbon $= 208.4$ kmol

Hydrogen atoms accounted in 100 kmol of dry output $= 153.0$ kmol

Unaccounted hydrogen atoms $= 208.4 - 153.0 = 55.4$ kmol

Unaccounted hydrogen atoms should have been present as water, formed by the oxidation of the hydrogen present in fuel as per the reaction

$$H_2 + 1/2\, O_2 \rightarrow H_2O$$

As the analysis is on dry basis, it has not been reported.

2 atoms of hydrogen are present in 1 molecule of water.

Water formed $= \dfrac{55.4}{2} = 27.7$ kmol

1 molecule of water contains 1 atom of oxygen. Oxygen atoms present in water have not been accounted for in the dry ouput from the burner.

Oxygen used in water formation $= 27.7$ kmol

Total amount of oxygen atoms fed to the burner = amount of oxygen atoms in dry output + amount of oxygen atoms in water formed

$= 34.2 + 27.7 = 61.9$ kmol

1 oxygen molecule contains 2 atoms.

Oxygen gas fed to the burner $= \dfrac{61.9}{2} = 30.95$ kmol

Methane gas fed to the reactor $= 52.1$ kmol

Unreacted methane in the output $= 5.2$ kmol

Methane reacted $= 52.1 - 5.2 = 46.9$ kmol

Conversion of methane $= \dfrac{46.9 \times 100}{52.1} = 90.02\%$

Acetylene formed on reaction of 46.9 kmol methane is 15 kmol.

$$\text{Yield of acetylene} = \frac{15 \times 100}{46.9} = 31.98\%$$

4F Acetaldehyde is oxidised over silica gel with the help of air. The mixture is passed over the catalyst at 387 K. The outgoing dry gases are found to contain 4.85% carbon dioxide, 8.65% acetaldehyde, 14.9% acetic acid, 2.55% oxygen and 69.05% nitrogen by volume on dry basis. For carrying out the dry analysis, water was first removed from the mixture. During water removal, some acetic acid was also removed by condensation. Calculate (i) the percentage conversion of acetaldehyde, (ii) the percentage yield of acetic acid, (iii) the ratio of air and acetaldehyde in the incoming feed, (iv) the percentage of acetic acid removed during water removal, and (v) the actual analysis of gases leaving the reactor.

Solution

Basis: The problem is similar to the previous problem. So 100 kmol of dry outgoing gases is the basis here also.

Acetaldehyde is oxidised by air and the reactions occurring are

$$CH_3CHO + 1/2\ O_2 \rightarrow CH_3COOH \tag{1}$$

$$CH_3CHO + 5/2\ O_2 \rightarrow 2CO_2 + 2H_2O \tag{2}$$

In the reactor, acetaldehyde is oxidised by air. The output from the reactor contains unreacted acetaldehyde, excess oxygen and inert nitrogen, and the products formed are acetic acid, carbon dioxide and water vapour. For carrying out dry analysis, the gaseous output is cooled. During this operation, water vapour is removed by condensation. It is given in the problem that some amount of acetic acid was also removed by condensation. Then the dry outgoing gases contain the remaining acetic acid only.

The source of nitrogen in the outgoing gases is air. Air contains 79mol% nitrogen and 21mol% oxygen. In the whole process, nitrogen is not affected. Knowing the amount of nitrogen and the amount of oxygen, the amount of air fed can be calculated. Amount of water vapour and acetic acid can be calculated by oxygen balance.

Nitrogen in the dry out going gases = 69.05 kmol.

Air contains 79mol% nitrogen.

$$\text{Amount of air fed} = \frac{69.05 \times 100}{79} = 87.41\ \text{kmol}$$

Air contains 21mol% oxygen.

Amount of oxygen from air $= 87.41 \times \frac{21}{100} = 18.36$ kmol

Unreacted oxygen present in the dry outgoing gases $= 2.55$ kmol

Reacted oxygen $= 15.81$ kmol

Oxygen is consumed in reaction (1) and reaction (2). In reaction (2), 1 mol of acetaldehyde reacts with 2.5 mol of oxygen, to give 2 mol of water and 2 mol of carbon dioxide.

Carbon dioxide in the dry outgoing gases $= 4.85$ kmol

Water formed $= 4.85$ kmol

Oxygen consumed in reaction (2) $= 4.85 \times \frac{2.5}{2.0} = 6.06$ kmol

Balance oxygen consumed in reaction (1) $= 9.75$ kmol

In reaction (1), each mole of oxygen consumed gives two moles of acetic acid.

Acetic acid actually formed $= 9.75 \times 2 = 19.5$ kmol

Acetic acid present in the dry outgoing gases $= 14.9$ kmol

Acetic acid removed during water removal = acetic acid formed − acetic acid detected $= 4.6$ kmol

Percentage of acetic acid removed during water removal $= \frac{4.6 \times 100}{19.5}$

$= 23.6\%$

In reaction (1), one mole of acetaldehyde gives one mole of acetic acid.

Acetaldehyde undergoing reaction (1) $= 19.5$ kmol

In reaction (2), one mole of acetaldehyde gives two moles of carbon dioxide.

Acetaldehyde undergoing reaction (2) $= \frac{4.85}{2} = 2.425$ kmol

Acetaldehyde reacted $= 19.5 + 2.425 = 21.925$ kmol

Acetaldehyde unreacted and present in outgoing gases $= 8.65$ kmol

Acetaldehyde fed to the reactor $= 21.925 + 8.65$

$= 30.575$ kmol

$$\text{Percentage conversion of acetaldehyde} = \frac{21.925 \times 100}{30.575} = 71.5\%$$

$$\text{Percentage yield of acetic acid} = \frac{19.5 \times 100}{21.925} = 88.94\%$$

$$\text{Acetaldehyde:air ratio} = 30.575 : 87.41 = 1 : 2.86$$

The actual outgoing gases contain water vapour in addition to the gases reported in the dry gas analysis given in the problem.

Table 4.3 gives the actual analysis of outgoing gases.

Table 4.3 Actual analysis of the outgoing gases

Constituent	Kilomoles (kmol)	Mole per cent
N_2	69.05	63.09
O_2	2.55	2.33
CH_3CH0	8.65	7.90
CH_3COOH	19.50	17.82
CO_2	4.85	4.43
Water vapour	4.85	4.43
	109.45	

4G Pure sulphur is burnt in air. In the process 20% excess of oxygen is used above that necessary for complete combustion of sulphur to sulphur trioxide, but only 30% sulphur burns to sulphur trioxide and the balance to sulphur dioxide. The gases are passed through a converter where conversion of sulphur dioxide to sulphur trioxide takes place. No new material is added to the converter. The gases leaving the converter have a mole fraction of 0.043 for oxygen. Calculate (i) the composition of the resulting mixture of gases in moles, and (ii) percentage ratio of moles of sulphur trioxide to moles of sulphur dioxide in the converter gases.

Solution

(i) *Basis*: 1 mole of sulphur burnt

The desired reaction is

$$S + 1/2\,O_2 \rightarrow SO_3$$

Theoretical oxygen required $= 1.5$ mol

20% excess oxygen is supplied.

$$\text{Actual oxygen supplied} = \frac{1.5 \times 120}{100} = 1.8\,\text{mol}$$

The source of oxygen is air. Air contains 79mol% nitrogen and 21mol% oxygen.

$$\text{Nitrogen from air} = \frac{1.8 \times 79}{21} = 6.77\,\text{mol}$$

The reactions occurring in the burner are:

$$S + O_2 \rightarrow SO_2$$

$$S + 1/2\,O_2 \rightarrow SO_3$$

30% sulphur burns to sulphur trioxide and the balance to sulphur dioxide.

So 0.3 mol sulphur trioxide and 0.7 mol sulphur dioxide are formed.

Oxygen consumption is 1.0 mol per mole for sulphur dioxide and 1.5 mol per mole for sulphur trioxide formation.

$$\text{Oxygen consumed} = (0.3 \times 1.5) + (0.7 \times 1.0) = 1.15\,\text{mol}$$

Composition of burner gases

Sulphur dioxide formed = 0.7 mol

Sulphur trioxide formed = 0.3 mol

Unreacted oxygen = 1.8 − 1.15 = 0.65 mol

Inert nitrogen present = 6.77 mol

The burner gases are fed to the converter. The reaction occurring in the converter is

$$SO_2 + 1/2\,O_2 \rightarrow SO_3$$

Let X mol of sulphur dioxide entering the converter get transformed to sulphur trioxide.

Composition of converter gases

Unreacted sulphur dioxide $= 0.7 - X$ mol

Sulphur trioxide formed $= 0.3 + X$ mol

Unreacted oxygen $= 0.65 - X/2$ mol

Inert nitrogen = 6.77 mol

Total number of moles of outgoing gases $= 8.42 - X/2$

$$\text{Mole fraction of oxygen} = \frac{\{0.65 - (X/2)\}}{\{8.42 - (X/2)\}} = 0.043 \text{ (given)}$$

So we get X $= 0.602$ mol

Table 4.4 gives the composition of the converter gases.

$$\text{Mole ratio of sulphur trioxide to sulphur dioxide} = \frac{0.902}{0.098} = 9.204$$

Table 4.4 Composition of converter gases

Constituent	Moles (mol)	Mole per cent
SO_2	0.098	1.2
SO_3	0.902	11.1
O_2	0.349	4.3
N_2	6.77	83.4
	8.119	

4H Iron pyrites are burnt in a sulphuric acid plant to give a burner gas which on Orsat analysis gives the result: sulphur dioxide 8.5%, oxygen 7.0% and nitrogen 84.5% (with traces of sulphur trioxide). Pyrites contain 46% by weight of sulphur. Solid refuse from the burner cinder contains 3wt% sulphur as ferric sulphate. The expected reactions are

$$4FeS_2 + 11O_2 \rightarrow 2Fe_2O_3 + 8SO_2$$

$$4FeS_2 + 15O_2 \rightarrow 2Fe_2O_3 + 8SO_3$$

$$Fe_2O_3 + 3SO_3 \rightarrow Fe_2(SO_4)_3$$

Calculate (i) the weight of cinder produced per 100 kg of pyrites burnt, (ii) the number of kilomoles of air used for 100 kg of pyrites burnt, and (iii) the number of kilomoles and composition of burner gas per 100 kg of pyrites burnt.

Solution

(i) *Basis*: 100 kg of pyrites burnt

Sulphur in 100 kg of pyrites $= 46$ kg

In pyrites (FeS_2), 2×32 kg of sulphur is in combination with 55.85 kg of iron. (Atomic weight of sulphur and iron are 32 and 55.85 respectively)

$$\text{Iron in 100 kg of pyrites} = \frac{46 \times 55.85}{2 \times 32} = 40.14 \text{ kg}$$

Balance is inerts.

Weight of inerts $= 100 - 46 - 40.14 = 13.86\,\text{kg}$

Cinder contains all the iron in pyrites as ferric oxide, inerts and sulphur trioxide. 2×55.85 kg iron results in 159.7 kg ferric oxide. Molecular weight of ferric oxide is 159.70.

Ferric oxide in cinder $= \dfrac{40.14 \times 159.70}{2 \times 55.85} = 57.39\,\text{kg}$

Ferric oxide + inerts in cinder $= 71.25\,\text{kg}$

Cinder contains 3wt% sulphur or $\dfrac{3 \times 80}{32} = 7.5$wt% sulphur trioxide

Sulphur trioxide-free cinder or 92.5% of cinder $= 71.25\,\text{kg}$

Actual cinder produced per 100 kg pyrites $= \dfrac{71.25 \times 100}{92.5} = 77.03\,\text{kg}$

(ii) *Basis*: The composition and rate of input are not precisely known. The problem has to be solved based on the output, even though composition of output is partially known. Orsat analysis of the output from the reactor does not include sulphur trioxide. However, the given data are sufficient to extract the unknown information. So 100 kmol sulphur trioxide-free burner gas is chosen as the basis.

Nitrogen in 100 kmol sulphur trioxide-free burner gas $= 84.5$ kmol

The source of nitrogen is air. Air contains 79mol% nitrogen and 21mol% oxygen. Using this, oxygen fed can be calculated

Oxygen fed to the burner $= 84.5 \times \dfrac{21}{79} = 22.46$ kmol

Unreacted oxygen in the burner gases $= 7.00$ kmol

Oxygen consumed $= 15.46$ kmol

Oxygen is consumed in reaction(1) and reaction (2). In reaction(1), 8 mol of sulphur dioxide are formed at the expense of 11 mol of oxygen.

Sulphur dioxide in 100 kmol sulphur trioxide-free burner gases $= 8.5$ kmol

Oxygen consumed in reaction (1) $= \dfrac{8.5 \times 11}{8} = 11.69$ kmol

Balance oxygen consumed in reaction (2) $= 3.77$ kmol

In reaction (2), 8 mol of sulphur trioxide are formed at the expense of 15 mol of oxygen.

Sulphur trioxide actually formed in the burner per 100 kmol sulphur trioxide-free burner gases $= \dfrac{3.77 \times 8}{15} = 2.01$ kmol

Total sulphur oxides in 100 kmol sulphur trioxide-free burner gas $= 8.5 + 2.01 = 10.51$ kmol

1 mol sulphur oxides requires 1 mol sulphur.

Sulphur required $= 10.51$ kmol

$= 10.51 \times 32 = 336.32$ kg

Source of sulphur is pyrites, and pyrites contain 46% by weight of sulphur.

Pyrites required $= \dfrac{336.32 \times 100}{46} = 731.13$ kg

Air fed for 731.13 kg pyrites = number of kilomoles of nitrogen and oxygen

$= 84.5 + 22.46 = 106.96$ kmol

Air fed for 100 kg pyrites $= \dfrac{106.96 \times 100}{731.13} = 14.63$ kmol

(iii) *Basis*: In the above two sections we have found out the actual amount of pyrites and air fed to the burner. We can use that as the basis, that is, 731.13 kg pyrites and 106.96 kmol air.

Sulphur trioxide formed in the burner (calculated in the previous section) has to be added to the composition of sulphur trioxide-free burner gas composition given in the problem. Table 4.5 gives the composition of gases formed.

Table 4.5 Composition of gases formed

Constituent	Kilomoles (kmol)
SO_2	8.5
SO_3	2.01
O_2	7.0
N_2	84.5

From this, some sulphur trioxide is lost to the cinder by absorption in ferric oxide.

Cinder formed per 100 kg pyrites $= 77.03$ kg

Cinder formed for 731.13 kg pyrites $= \dfrac{77.03 \times 731.13}{100} = 563.19$ kg

As worked out in the first section, 7.5% of cinder is sulphur trioxide.

$$\text{Sulphur trioxide lost in cinder} = \frac{563.19 \times 7.5}{100} = 42.23\,\text{kg} = \frac{42.23}{80} = 0.53\,\text{kmol}$$

Composition of the gases leaving the burner is calculated, by subtracting sulphur trioxide lost in the cinder from the burner gases formed and is given in Table 4.6.

Table 4.6 Composition of gases leaving the burner

Constituent	Amount (kmol)	Mole per cent
SO_2	85.00	8.38
SO_3	2.01 − 0.53 = 1.48	1.46
O_2	7.0	6.90
N_2	84.5	83.26
	101.48	

$$\text{Amount of burner gas for 100 kg of pyrites} = \frac{101.48 \times 100}{731.13} = 13.88\,\text{kmol}$$

4I Brine on electrolysis produces sodium hydroxide, hydrogen and chlorine as per the reaction

$$2NaCl + 2H_2O \rightarrow 2NaOH + H_2 + Cl_2$$

A cell is being fed with a brine solution (specific gravity:1.20) containing 25% sodium chloride and produces a caustic lye (specific gravity:1.14) analysing 16% sodium chloride and 6% sodium hydroxide. The lye is concentrated to 50% sodium hydroxide during which process sodium chloride crystallises out and is recycled. If the feed to the cell is $3 \times 10^{-2}\,m^3\,h^{-1}$, (i) How much of sodium chloride is recycled? (ii) How much of 50% sodium hydroxide is produced? (iii) How much of chlorine and hydrogen are produced in cubic metres at STP?

Solution

Basis: 1 hour operation

For sodium, there is one input (brine fed in) and only one output (lye produced). The problem can be solved by working out the sodium balance.

$$\text{Brine fed to the cell} = 30\,\text{L} = 30 \times 1.2 = 36\,\text{kg}\ (\text{density} = 1.2\,\text{kg L}^{-1})$$

Brine contains 25% by weight of sodium chloride.

$$\text{Sodium chloride in the feed} = \frac{36 \times 25}{100} = 9\,\text{kg}$$

58.46 g of sodium chloride contain 23 g of sodium. (58.47 is the molecular weight of sodium chloride and 23 is the atomic weight of sodium.)

Sodium fed to the cell $= \dfrac{9\,000 \times 23}{58.46} = 3\,540.88$ g

1 litre lye weighs 1.14 kg (density = 1.14 kg L^{-1}).

Lye contains 16% sodium chloride and 6% sodium hydroxide by weight. 58.46 g of sodium chloride contains 23 g of sodium. 40 g of sodium hydroxide contains 23 g of sodium. (Molecular weight of sodium chloride and sodium hydroxide are 58.46 and 40 respectively, and the atomic weight of sodium is 23.)

Sodium in 1.14 kg lye = sodium in sodium chloride + sodium in sodium hydroxide

$$= 1.14 \times \frac{16}{100} \times \frac{23}{58.46} + 1.14 \times \frac{6}{100} \times \frac{23}{40}$$

$$= 0.111\,09 \text{ kg}$$

Lye to be produced for 3 540.88 g of sodium $= \dfrac{3\,540.88}{0.111\,09} \times 1.14 = 36.336$ kg

Sodium hydroxide in the lye produced $= \dfrac{36.336 \times 6}{100} = 2.18 \text{ kg} = \dfrac{2.18}{40}$ kmol

6% sodium hydroxide is concentrated to 50% sodium hydroxide.

50% sodium hydroxide produced $= \dfrac{2.18 \times 100}{50} = 4.36 \text{ kg h}^{-1}$

Sodium chloride in the lye (16 % of lye) $= \dfrac{36.336 \times 16}{100} = 5.813$ kg

This sodium chloride is recycled.

Sodium chloride recycled $= 5.813 \text{ kg h}^{-1}$

1 mol of sodium hydroxide is accompanied by 0.5 mol each of hydrogen and chlorine.

Chlorine or hydrogen produced $= 0.5 \times \dfrac{2.18}{40}$ kmol

$$= 0.5 \times \frac{2.18}{40} \times \frac{0.082\,05 \times 298}{1}$$

$$= 0.666\,3 \text{ m}^3 \text{ h}^{-1} \text{ at STP}$$

4J A certain gaseous hydrocarbon is known to contain less than five carbons. This gas is burnt with the theoretically required oxygen and complete combustion takes place. Volume of the reactants is 600 ml, while that of the products (also all gases) is 700 ml under identical conditions. What is the hydrocarbon?

Solution

Let the hydrocarbon be C_xH_y

The combustibles here are carbon and hydrogen. The combustion reactions are

$$C + O_2 \rightarrow CO_2$$

$$H + 1/4\, O_2 \rightarrow 1/2\, H_2O$$

$$C_xH_Y + \{x + (y/4)\}O_2 \rightarrow xCO_2 + (y/2)H_2O$$

$$\frac{\text{Reactant volume}}{\text{Product volume}} = \frac{[1 + \{x + (y/4)\}]}{\{1x + (y/2)\}} = \frac{600}{700}$$

This reduces to $4x + 28 = 5y$

$y = (4x + 28)/5$

For the hydrocarbon, the values of x and y should be whole numbers.

When	$x = 1$	$y = 6.4$
	$x = 2$	$y = 7.2$
	$x = 3$	$y = 8$
	$x = 4$	$y = 8.8$

As $x < 5$ we need not go further.

So the hydrocarbon should be C_3H_8. Its combustion reaction is

$$C_3H_8 + 5O_2 \rightarrow 3CO_2 + 4H_2O$$

6 mol/6 vol of reactants give 7 mol/7 vol of products.

4K Hydrofluoric acid (HF) is manufactured by treating fluorspar with sulphuric acid. A sample of fluorspar contains 75wt% calcium fluoride and the balance is inerts. Aqueous 80wt% sulphuric acid is used. To ensure completion of reaction, 20% excess of acid is used. HF formed volatilises from the reaction mass. 5% of HF formed remains in the residue left in the reactor. Based on a charge of 100 kg fluorspar, calculate the weight and composition of the residue obtained.

Solution

Basis: 100 kg of fluorspar fed to the reactor

Calcium fluoride in 100 kg fluorspar = 75 kg

Inerts = 25 kg

Fluorspar is treated with sulphuric acid. The reaction occurring is given below. The molecular weights of the components are also indicated.

$$\underset{78}{CF_2} + \underset{98}{H_2SO_4} \rightarrow \underset{136}{CaSO_4} + \underset{40}{2HF}$$

78 kg of calcium fluoride require 98 kg of sulphuric acid.

Theoretical weight of 100% pure sulphuric acid required $= \dfrac{98 \times 75}{78} = 94.23$ kg

20% excess sulphuricacid is used in the process.

Sulphuric acid supplied $= \dfrac{94.23 \times 120}{100} = 113.08$ kg

Only aqueous 80% sulphuric acid is used.

Water in aqueous sulphuric acid $= \dfrac{113.08 \times 20}{80} = 28.27$ kg

Hydrofluoric acid formed $= \dfrac{40 \times 75}{78} = 38.46$ kg

5% of HF formed stays in the residue, while the rest volatilises off.

Hydrofluoric acid in the residue $= \dfrac{38.46 \times 5}{100} = 1.93$ kg

Calcium sulphate formed $= \dfrac{136 \times 75}{78} = 130.77$ kg

Unreacted sulphuric acid $= \begin{matrix}\text{sulphuric acid} \\ \text{supplied}\end{matrix} - \begin{matrix}\text{theoretical requirement} \\ \text{of sulphuric acid}\end{matrix}$

$= 113.08 - 94.23$

$= 18.85$ kg

The residue consists of calcium sulphate formed, remaining hydrofluoric acid, water in aqueous sulphuric acid, excess sulphuric acid and inerts present in the ore. The composition of the residue is given in Table 4.7.

Table 4.7 Composition of residue

Constituent	Weight (kg)	Weight per cent
$CaSO_4$	130.77	63.85
HF	1.93	0.94
H_2O	28.27	13.80
H_2SO_4	18.85	09.20
Inerts	25.00	12.21
	204.82	

4L Limestone containing 82% of calcium carbonate, 13% of magnesium carbonate and 5% of silica is calcined by heating to 1 273 K to drive off carbon dioxide gas and form calcium oxide and magnesium oxide. Calcination of calcium carbonate is 95% complete and that of magnesium carbonate is 90% complete. (i) What is the analysis of lime obtained by calcinations? (ii) What is the weight of carbon dioxide obtained by the calcination of 1000 kg of limestone?

Solution

Basis: 1000 kg of limestone

Limestone is a solid and hence the valid assumption for analysis is weight per cent.

Calcium carbonate in 1000 kg limestone $= \dfrac{1000 \times 82}{100} = 820\,\text{kg}$

Magnesium carbonate in 1000 kg limestone $= \dfrac{1000 \times 13}{100} = 130\,\text{kg}$

Silica in 1000 kg limestone (by difference) $= 50\,\text{kg}$

Calcination reaction of calcium carbonate is given below. It is 95% complete.

$$\underset{100}{CaCO_3} \longrightarrow \underset{56}{CaO} + \underset{44}{CO_2}$$

Calcium oxide formed $= \dfrac{56 \times 820 \times 0.95}{100} = 436.24\,\text{kg}$

Carbon dioxide formed $= \dfrac{44 \times 820 \times 0.95}{100} = 342.76\,\text{kg}$

Unreacted calcium carbonate $= 820 \times 0.05 = 41.5\,\text{kg}$

Calcination reaction of magnesium carbonate is given below. It is 95% complete.

$$\underset{84.32}{MgCO_3} \longrightarrow \underset{40.32}{MgO} + \underset{44}{CO_2}$$

$$\text{Magnesium oxide formed} = \frac{40.32 \times 130 \times 0.90}{84.32} = 55.95\,\text{kg}$$

$$\text{Carbon dioxide formed} = \frac{44 \times 130 \times 0.90}{84.32} = 61.05\,\text{kg}$$

$$\text{Unreacted magnesium carbonate} = \frac{130 \times 10}{100} = 13\,\text{kg}$$

$$\text{Total carbon dioxide formed} = \begin{array}{c}\text{carbon dioxide from}\\ \text{calcium carbonate}\end{array} + \begin{array}{c}\text{carbon dioxide from}\\ \text{magnesium carbonate}\end{array}$$

$$= 342.76 + 61.05 = 403.81\,\text{kg}$$

Analysis of lime produced is given in Table 4.8.

Table 4.8 Analysis of lime

Constituent	Weight (kg)	Weight per cent
$CaCO_3$	41.50	6.88
CaO	436.24	73.17
$MgCO_3$	13.00	2.18
MgO	55.95	9.38
SiO_2	50.00	8.39
	596.69	

4M Sulphur dioxide reacts with pure oxygen to form sulphur trioxide. The reaction is carried out with 100% excess oxygen. But due to the low temperature and pressure, completion of reaction is only 60%. Calculate (i) the mole fraction of sulphur dioxide, oxygen and sulphur trioxide in reactants and products, (ii) the weight fractions of these gases in the reactants and products, (iii) the average molecular weight of reactants and products, (iv) the partial pressure of the products if the total pressure is 2 atm, and (v) the volume of the gases obtained from 100 kg sulphur dioxide, if the products were cooled to 293 K and pressure reduced to 735 mm Hg pressure.

Solution

Basis: 1 kmol sulphur dioxide fed to the reactor

The reaction taking place is

$$SO_2 + 1/2\,O_2 \rightarrow SO_3$$

Theoretical oxygen required = 0.5 kmol

100% excess oxygen is fed.

Oxygen actually fed = 1 kmol

Table 4.9 gives the analysis of the reactants.

Table 4.9 Analysis of the reactants

Constituent	Kilomoles (kmol)	Mole fraction	Molecular weight	Weight (kg)	Weight fraction
SO_2	1.0	0.5	64	64	0.666
O_2	1.0	0.5	32	32	0.334
	2.0			96	

2 kmol of reactants weigh 96 kg and hence average molecular weight is 48.

Degree of completion is 60%. Of the 1.0 kmol of sulphur dioxide fed, 60%, that is 0.6 kmol, forms products. As per the reaction given above, 1 mol sulphur dioxide reacts with 0.5 mol oxygen to give 1 mol sulphur trioxide.

So 0.6 kmol sulphur dioxide consumes 0.3 kmol of oxygen to form 0.6 kmol of sulphur trioxide.

Unreacted sulphur dioxide = sulphur dioxide fed − sulphur dioxide reacted

= 1.0 − 0.6 = 0.4 kmol

Unreacted oxygen = oxygen fed − oxygen consumed

= 0.5 − 0.3 = 0.2 kmol

Sulphur trioxide formed = 0.6 kmol

Table 4.10 gives the analysis of the products.

Table 4.10 Analysis of the products

Constituent	Kilomoles (kmol)	Mole fraction	Molecular weight	Weight (kg)	Weight fraction
SO_2	0.4	0.235	64	25.6	0.267
O_2	0.7	0.412	32	22.4	0.233
SO_3	0.6	0.353	80	48	0.5
	1.7			96	

1.7 kmol weigh 96 kg and hence average molecular weight is 56.47.

Partial pressure is given by mole fraction × total pressure.

Partial pressure of sulphur dioxide $= 0.235 \times 2 = 0.470$ atm

Partial pressure of sulphur trioxide $= 0.353 \times 2 = 0.706$ atm

Partial pressure of oxygen $= 0.412 \times 2 = 0.824$ atm

100 kg sulphur dioxide $= \dfrac{100}{64} = 1.562\,5$ kmol

1 kmol of sulphur dioxide gives 1.7 kmol of product gases.

1.562 5 kmol sulphur dioxide will give $= 1.7 \times 1.562\,5$ kmol or 2.656 25 kmol

Volume of the product gases at 293 K and 735 mm Hg pressure $= \dfrac{2.6562\,5 \times 0.082\,05 \times 293 \times 760}{735} = 66.03\ \text{m}^3$

4N In the Kroll process for making titanium, titanium oxide mixed with carbon and chlorine is passed through the reactor. Titanium tetrachloride vapours distill off. They are condensed and reduced to titanium sponge by reaction with magnesium metal in an atmosphere of argon. The reactions are

$$TiO_2 + C + 2Cl_2 \rightarrow TiCl_4 + CO_2$$

$$TiCl_4 + 2Mg \rightarrow Ti + 2MgCl_2$$

Calculate the theoretical weights of chlorine, carbon and magnesium required to produce one tonne of titanium, assuming that the reaction is complete. Also find the weight of the titanium tetrachloride intermediate.

Solution

Basis: 1000 kg of titanium

The overall reaction is

$$TiO_2 + C + 2Cl_2 + 2Mg \rightarrow Ti + 2MgCl_2 + CO_2$$

Atomic weights of titanium, magnesium, carbon and molecular weight of chlorine are 47.9, 24.32, 12 and 70.92 respectively.

Weight of magnesium required by 47.9 kg of titanium $= 2 \times 24.32 = 48.64$ kg

Weight of magnesium required by 1000 kg of titanium $= \dfrac{48.64 \times 1000}{47.9} = 1\,015.45$ kg

Carbon required by 47.9 kg of titanium $= 12$ kg

Carbon required by 1000 kg of titanium $= \dfrac{12 \times 1000}{47.9} = 250.52$ kg

Chlorine required by 47.9 kg of titanium $= 2 \times 70.92$

$= 141.94$ kg

Chlorine required by 1000 kg of titanium $= \dfrac{141.94 \times 1000}{47.9}$

$= 2\,961.17$ kg

Weight of titanium tetrachloride intermediate formed = weight of titanium + weight of chlorine

$= 1000 + 2\,961.17 = 3\,961.17$ kg

40 Monochlorobenzene is to be manufactured by reacting benzene with chlorine.

$$C_6H_6 + Cl_2 \rightarrow C_6H_5Cl + HCl$$

It is known that 80% of chlorine reacting goes to produce chlorobenzene and the other 20% forms dichlorobenzene.

$$C_6H_6 + 2Cl_2 \rightarrow C_6H_4Cl_2 + 2HCl$$

How much benzene and chlorine are needed for producing 1000 kg of chlorobenzene? How much hydrogen chloride will be produced? What percentage of chlorine is converted to monochlorobenzene?

Solution

Basis: 1000 kg chlorobenzene produced

Molecular weight of chlorobenzene is 112.5.

Amount of monochlorobenzene in 1000 kg $= \dfrac{1000}{112.55} = 8.89$ kmol

For this amount of monochlorobenzene an equal amount of benzene and chlorine on a molar scale are required and the same amount of hydrogen chloride is produced. When 80% of chlorine goes to chlorobenzene, 20% of chlorine goes to dichlorobenzene.

Chlorine utilised for forming dichlorobenzene $= \dfrac{8.89 \times 20}{80} = 2.22$ kmol

2 mol of chlorine are required for 1 mol of dichlorobenzene.

Dichlorobenzene formed $= \dfrac{2.222\,5}{2} = 1.11$ kmol

1 mol of benzene is required for every mole of monochlorobenzene and dichlorobenzene.

1.11 kmol of benzene is required for 1.11 kmol of dichlorobenzene.

Total benzene required $= 8.89 + 1.11 = 10.00\,\text{kmol}$

Chlorine required $= 8.89 + 2.22 = 11.11\,\text{kmol}$

1 mol of hydrogen chloride is produced for each mole of chlorine consumed.

Hydrogen chloride produced $= 11.11\,\text{kmol}$

Of the chlorine undergoing the reaction,

$$C_6H_6 + Cl_2 \rightarrow C_6H_5Cl + HCl$$

half goes towards forming chlorobenzene and the balance towards forming hydrogen chloride.

So percentage of chlorine converted to chlorobenzene $= 80 \div 2 = 40\%$

4P Commercial nitric acid is produced by the catalytic air oxidation of ammonia, according to the following equation

$$4NH_3 + 5O_2 \rightarrow 4NO + 6H_2O$$

This reaction goes to 95% completion. It is followed by the reaction,

$$4NO + 3O_2 + 2H_2O \rightarrow 4HNO_3$$

This reaction goes to 98% completion. Calculate the amount of ammonia and water required for producing one tonne of 65% nitric acid.

Solution

Basis: 1000 kg of 65% nitric acid

Nitric acid in 1000 kg of 65% nitric acid $= 650\,\text{kg} = \dfrac{650}{63} = 10.32\,\text{kmol}$

Molecular weight of nitric acid is 63. The reactions involved in the production of nitric acid are

$$4NH_3 + 5O_2 \rightarrow 4NO + 6H_2O \quad (1)$$

$$4NO + 3O_2 + 2H_2O \rightarrow 4HNO_3 \quad (2)$$

As per reaction (2), 1 mol of nitric oxide gives 1 mol of nitric acid.

Theoretical nitric oxide required $= 10.32\,\text{kmol}$

But reaction (2) goes only to 98% completion.

Actual nitric oxide required $= \dfrac{10.32}{0.98} = 10.53\,\text{kmol}$

As per reaction (2), conversion of nitric oxide to nitric acid requires 0.5 mol of water per mole of nitric acid.

Water required $= \dfrac{10.53}{2} = 5.265\,\text{kmol} = 5.265 \times 18 = 94.77\,\text{kg}$

Nitric oxide is produced by the oxidation of ammonia, as per reaction(1).

1 mol of ammonia is required for the production of 1 mol of nitric oxide.

Ammonia required $= 10.53\,\text{kmol}$

But conversion of ammonia to nitric oxide is only 95% complete.

Actual ammonia required $= \dfrac{10.53 \times 100}{95} = 11.08\,\text{kmol}$

$= 11.08 \times 17 = 188.43\,\text{kg}$

In the conversion of ammonia to nitric oxide, as per reaction (1) each mole of ammonia produces 1.5 mol of water.

Water formed in the conversion of ammonia to nitric oxide $= 11.08 \times 1.5$

$= 16.62\,\text{kmol} = 299.16\,\text{kg}$

Water required for diluting nitric acid formed to 65% aqueous nitric acid $= 350\,\text{kg}$

Total water required = water for dilution + water for conversion of nitric oxide to nitric acid − water produced in conversion of ammonia to nitric oxide

$= 350 + 94.77 - 299.16 = 145.61\,\text{kg}$

ELECTROCHEMICAL REACTIONS

Reactions where an electric current is one of the reactants are known as electrochemical reactions. Chlor-alkali industry, aluminum industry, copper industry and electroplating are some examples of electrochemical industries.

In electrochemical reactions, the amount of product formed depends on the quantity of current utilized. The *quantity of current* is obtained by multiplying the current applied by the duration of application of current. When one ampere current is passed for one second, the quantity of current passed is one ampere second (1 A s) or one coulomb (1 C).

One faraday (1 F) or 96 500 C of current liberate one gram equivalent of material at the electrodes. This is known as Faraday's First law of electrolysis. 1 F of current gives 40 g of sodium hydroxide or liberates 8 g of oxygen or 1.008 g of hydrogen or 35.46 g of chlorine—all at 100% current efficiency—or the theoretical yield.

Not all the current passed produces the desired reaction, that is, the process is not 100% efficient. The actual production depends on the current efficiency of the process. *Current efficiency* is defined as the percentage ratio of actual production to theoretical production. Theoretical production is calculated based on Faraday's First law, while actual production is what we actually obtain.

The product of quantity of current passed and cell voltage gives *power consumption*. Based on this, we can calculate *specific power consumption* for any electrochemical product.

When current is passed through an electrolytic cell, the quantity of current passed through both the electrodes—anode and cathode—is the same. It may also be noted that current efficiency is the same for both the electrodes.

ILLUSTRATIONS

4Q When 10 A current was passed for 19 300 seconds through a silver plating unit, 200 g of Ag were deposited on the cathode. Calculate the amount of oxygen evolved at the anode and the current efficiency.

Solution

Faraday's First law states that 1 F or 96 500 C of current liberate one gram equivalent of material at the electrodes.

Equivalent weight of silver is 107.88. Therefore 96 500 C give 107.88 g of silver.

$$(19\,300 \times 10)\text{ C will give } \frac{107.88 \times 19\,300 \times 10}{96\,500} \text{ or } 215.76 \text{ g of silver.}$$

This is the theoretical production.

Actual production as given in the problem = 200 g

Current efficiency = percentage ratio of actual production to theoretical production

$$= \frac{200 \times 100}{215.76} = 92.7\%$$

Current efficiency is the same for anode and cathode. In silver plating, the cathodic reaction is silver deposition and the anodic reaction is oxygen evolution.

96 500 C current should theoretically give 8 g of oxygen, where 8 is the equivalent weight of oxygen.

Theoretical production of oxygen for (19 300 × 10) C current $= \dfrac{8 \times 19\,300 \times 10}{96\,500}$ g

Actual amount of oxygen evolved at the anode = Theoretical production × current efficiency

$$= \frac{8 \times 19\,300 \times 10}{96\,500} \times \frac{92.7}{100} = 14.832 \text{ g}$$

4R An industrial electrochemical hydrogen generator works at 95% current efficiency. Find the current required for producing 1 m^3 (at NTP) of hydrogen per hour. If the cell voltage is 2.5 V, calculate specific power consumption in terms of kWh kg^{-1} of hydrogen.

Solution

Basis: 1 hour operation.

Hydrogen produced in one hour = 1 m^3 at NTP

Amount of hydrogen produced in one hour $= \dfrac{1}{22.4}$ mol

Weight of hydrogen produced in one hour $= \dfrac{1 \times 2.016}{22.4} = 0.09 \text{ kg} = 90 \text{ g}$

Faraday's First law states that 1 F or 96 500 C of current liberate 1 gram equivalent of material at the electrodes.

Equivalent weight of hydrogen is 1.008, and 96 500 C gives 1.008 g of hydrogen.

Quantity of current required for the production of 90 g of hydrogen $= \dfrac{96\,500 \times 90}{1.008}$ C

Duration of current = 1 hour = 3 600 s

Current to be passed $= \dfrac{96\,500 \times 90}{1.008 \times 3\,600} = 2\,394$ A

This is at 100% current efficiency.

Current required, if current efficiency is 95% = 2 394 ÷ 0.95 = 2 520 A

Power consumed for the production of in one hour = 2 520 A × 1 h × 2.5 V = 6 300 Wh = 6.3 kWh

Power consumed 0.09 kg hydrogen, = 6.3 kWh

Specific power consumption = 6.3 ÷ 0.09 = 70 kWh kg^{-1}

4S A diaphragm cell for the production of sodium hydroxide, chlorine and hydrogen, operates at a load of 20 000 A. Aqueous sodium chloride solution fed to the cell contains 26% sodium chloride. The cell effluent contains 12% sodium hydroxide and has a salt to caustic ratio of 1.4. If the cell produces 680 kg the sodium hydroxide per day, find (i) the current efficiency of the process, (ii) the amount of chlorine and hydrogen produced, and (iii) the evaporation loss from the cell.

Solution

Basis: 1 day operation

(i) Coulumbs of current passed in one day = (20 000 × 24 × 3 600) C

Faraday's First law states that 1 F or 96 500 C of current liberates one gram equivalent of material at the electrodes.

Equivalent weight of sodium hydroxide is 40. Therefore, 96 500 C give 40 g of sodium hydroxide.

Theoretical production of sodium hydroxide in one day $= \dfrac{20\,000 \times 24 \times 3\,600 \times 40}{96\,500 \times 1000}$

$= 716.3$ kg

Actual production of sodium hydroxide in one day is only 680 kg.

Current efficiency is percentage ratio of actual production to theoretical production.

Current efficiency of the process $= \dfrac{680 \times 100}{716.3} = 94.94\%$

(ii) Sodium hydroxide produced $= 680 \text{ kg} = 680 \div 40 = 17$ kmol

Overall reaction is

$$NaCl + H_2O \rightarrow NaOH + 1/2\, H_2 + 1/2\, Cl_2$$

1 mol of sodium hydroxide is accompanied by 0.5 mol each of hydrogen and chlorine.

Chlorine or hydrogen produced $= \dfrac{17}{2} = 8.5$ kmol

$= 8.5 \times 22.4 = 190.4\ m^3$ at NTP per day

(iii) Salt to caustic soda ratio in cell liquor = 1.4

Sodium chloride in cell liquor = amount of sodium hydroxide × 1.4

$= 680 \times 1.4 = 952$ kg

This is unreacted sodium chloride. From the amount of sodium hydroxide in the cell liquor, reacted sodium chloride can be calculated.

40 g of sodium hydroxide is obtained from 58.46 g of sodium chloride.

Reacted sodium chloride $= \dfrac{680 \times 58.46}{40} = 993.8\text{ kg}$

Total sodium chloride in the feed is the sum of reacted and unreacted sodium chloride $= 993.8 + 952 = 1\,945.8\text{ kg}$

Sodium chloride forms 26% of the feed and the balance 74% is water.

Water in the feed $= \dfrac{1\,945.8 \times 74}{26} = 5\,538\text{ kg}$

Cell liquor contains 12% sodium hydroxide. Amount of sodium hydroxide in cell liquor is 680 kg.

Weight of cell liquor $= \dfrac{680 \times 100}{12} = 5\,666.7\text{ kg}$

Cell liquor contains sodium chloride, sodium hydroxide and water.

Water in cell liquor $= 5\,666.7 - \text{weight of sodium hydroxide} - \text{weight of sodium chloride}$

$= 5\,666.7 - 680 - 952 = 4\,034.7\text{ kg}$

40 g of sodium hydroxide is obtained by the consumption of 18 g of water.

Water consumed for the formation of 680 kg of sodium hydroxide $= \dfrac{680 \times 18}{40} = 306\text{ kg}$

Evaporation loss $= \text{water in feed} - \text{water in cell liquor} - \text{water consumed in sodium hydroxide formation.}$

$= 5\,538 - 4\,034.7 - 306 = 1\,197.3\text{ kg per day}$

4T Feed to a Hooker-S cell is 16.92 kg of sodium chloride per hour in an aqueous solution. Discharge from the cathode chamber on analysis shows 130 g sodium chloride and 135 g sodium hydroxide per 1000 g water. If the cell operates at 5 100 A and 3.4 V, compute (i) current efficiency, and (ii) production of sodium hydroxide and chlorine in kg kWh^{-1}.

Solution

Basis: 1 day operation

Sodium chloride fed to the cell in 1 hour $= 16.92\text{ kg}$

Sodium chloride is converted to sodium hydroxide by the reaction,

$$NaCl + H_2O \rightarrow NaOH + 1/2\, H_2 + 1/2\, Cl_2$$

The output from the cell contains 130 g sodium chloride and 135 g sodium hydroxide per 1000 g water. From the amount of sodium hydroxide in cell liquor, reacted sodium chloride can be calculated.

40 g of sodium hydroxide is obtained from 58.46 g of sodium chloride.

135 g of sodium hydroxide is obtained from $\dfrac{135 \times 58.46}{40}$ or 197.2 g sodium chloride.

Reacted sodium chloride $= 197.2$ g

Unreacted sodium chloride $= 130$ g

Total sodium chloride $= 327.2$ g

For 327.2 g of sodium chloride, sodium hydroxide formed is 135 g

For 16.92 kg sodium chloide, sodium hydroxide formed $= \dfrac{135 \times 16.92}{327.2}$

$= 6.98\ \text{kg h}^{-1}$

The product of quantity of current passed and cell voltage gives power consumption.

Power consumption in one hour $= 5.1\ \text{kA} \times 3.4\ \text{V} \times 1\ \text{h} = 17.34\ \text{kWh}$

Sodium hydroxide production in kg kWh^{-1} $= 6.98 \div 17.34 = 0.402\,6\ \text{kg kWh}^{-1}$

40 g sodium hydroxide is accompanied by 35.46 g of chlorine.

Chlorine produced $= \dfrac{0.402\,6 \times 35.46}{40} = 0.356\,9\ \text{kg kWh}^{-1}$

Coulombs of current passed in one hour $= (5\,100 \times 3\,600)$ C

Faraday's First law states that 1 F or 96 500 C of current liberates one gram equivalent of material at the electrodes.

Equivalent weight of sodium hydroxide is 40. Therefore 96 500 C gives 40 g of sodium hydroxide.

Theoretical production of sodium hydroxide in one hour $= \dfrac{5\,100 \times 3\,600 \times 40}{96\,500 \times 1000}$

$= 7.612$ kg

Actual production of sodium hydroxide in one hour is 6.98 kg.

Current efficiency is percentage ratio of actual production to theoretical production.

$$\text{Current efficiency of the process} = \frac{6.98 \times 100}{7.612} = 91.7\%$$

RECYCLE, BYPASS AND PURGE CALCULATIONS

Recycle

Processes involving recycling of part of the product are frequently used in the chemical industry. Based on economic considerations, it is a common practice to allow a reaction to proceed to partial completion. The unreacted material is then separated from the product and recycled. Often an excess of one component may be used to drive the reaction to completion and the excess is recovered and reused. In drying operations, the humidity of air is controlled by recirculating some of the wet air which leaves the drier. In fractionating columns, part of the distillate is refluxed through the column to enrich the product. In ammonia synthesis, the gas mixture leaving the converter is recycled back to the converter after recovery of the ammonia formed. There is recycle in every stage of petroleum processing.

Recycle stream is a stream which transfers part of the material coming out of a unit or a reactor back to the inlet of the same unit. The schematic representation of a typical recycle operation is given in Figure 4.1.

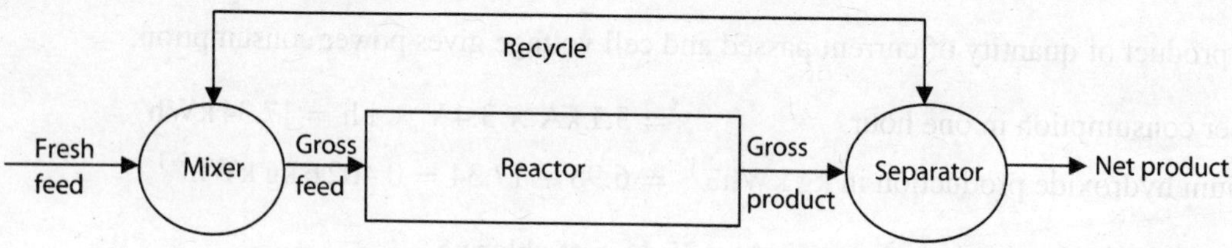

Fig. 4.1 Schematic representation of a typical recycle operation

The feed to the process is called *gross feed* and it is made up of fresh feed and recycled material. Gross product leaving the process is separated into the net product and the recycle stream. This separation may take different forms. When air is recirculated to a drier, only a flow proportionator is needed; then the compositions of the recycle stream and the net product are the same. In other cases, more complicated separation processes such as distillation or crystallisation are used. Such separations usually give a recycle stream with composition different from that of the product.

The procedure for solving recycle problems depends on the information available and also on what happens in the process and the separator. Mostly overall balance and balance over the mixer or separator will be adequate to solve the problem. Recycle ratio is defined as the ratio of recycle to fresh feed. This may be in mass or molar units.

Bypass stream

Bypassing a fluid stream by splitting it into two parallel streams is often practised when accurate control in concentration is desired. For example, in the conditioning of air with a single stationary bed of silica gel, to a lower humidity, a more nearly uniform humidity is obtained in the final air mixture if part of the air is dehumidified to a lower moisture content and then mixed with the unconditioned air to produce the desired mixture.

Bypass stream is a stream that skips one or more earlier stages of the process and goes directly to a later stage. Schematic representation of a typical bypass operation is given in Figure 4.2.

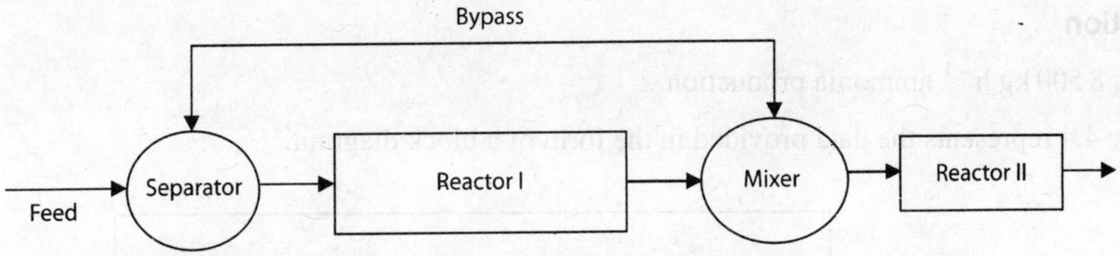

Fig. 4.2 Schematic representation of a typical bypass operation

Purge stream

One major drawback with the recycling of fluid streams is the gradual accumulation of inerts or impurities in the recycled stream. If some provision is not made for removing such impurities, they will gradually accumulate and the process will come to a stop. Bleeding off a fraction of the recycle stream can solve this problem. This type of *bleeding off* or *purging* is practised in the electrolytic recovery of copper to control the accumulation of impurities like nickel sulphate in the electrolyte, and in the synthesis of ammonia to limit the argon content in the synthesis gas within the tolerance limit.

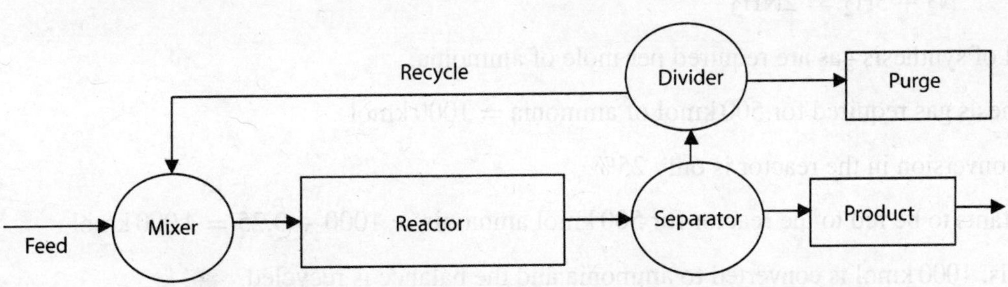

Fig. 4.3 Schematic representation of a typical purge operation

Purge stream is a stream bled off to remove an accumulation of inerts or unwanted material that might otherwise build up in the recycle stream. Schematic representation of a typical bypass operation is given in Figure 4.3.

ILLUSTRATIONS

4U An ammonia converter is fed with a 1:3 mixture of nitrogen and hydrogen. Conversion of ammonia is 25%, based on nitrogen. Ammonia formed is separated by liquefaction and the unconverted reactants are recycled back to the converter. Find (i) the volume of fresh feed, and (ii) the volume of recycle required to produce 8 500 kg h^{-1} of ammonia. Fresh feed and recycle are at 873 K and 2.5 atm gauge pressure.

Solution

Basis: 8 500 kg h^{-1} ammonia production

Figure 4.4 represents the data provided in the form of a block diagram.

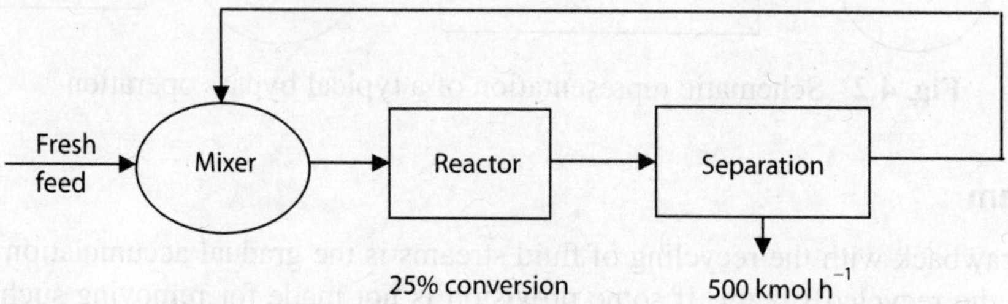

Fig. 4.4 Ammonia synthesis involving recycle

Ammonia produced = 8 500 ÷ 17 = 500 kmol

The reaction occurring in the reactor is

$$N_2 + 3H_2 \rightleftharpoons 2NH_3$$

2 mol of synthesis gas are required per mole of ammonia.

Synthesis gas required for 500 kmol of ammonia = 1000 kmol

But conversion in the reactor is only 25%.

Reactants to be fed to the reactor for 500 kmol ammonia = 1000 ÷ 0.25 = 4 000 kmol

Of this, 1000 kmol is converted to ammonia and the balance is recycled.

Amount of recycle = 4 000 − 1000 = 3 000 kmol

For the process to continue, 4 000 kmol is required, which is obtained by mixing recycle with fresh feed.

So fresh feed required = 4 000 − 3 000 = 1000 kmol

2.5 atm gauge pressure = 3.5 atm absolute pressure

Volume of fresh feed at 823 K and 3.5 atm pressure $= \dfrac{nRT}{P}$

$$= \frac{1000 \times 0.08205 \times 823}{3.5} = 19\,293.47 \text{ m}^3$$

Volume of recycle stream at 823 K and 3.5 atm $= \dfrac{3\,000 \times 0.08205 \times 823}{3.5}$

$$= 57\,880.41 \text{ m}^3$$

4V A solid material with 15wt% water is dried to 7wt% water under the following conditions. Fresh air is mixed with recycled air and blown over the solid. Fresh air contains 0.01 kg moisture per kilogram of dry air and recycled air which is part of the air leaving the drier contains 0.1 kg moisture per kilogram of dry air. Mixed air entering the drier contains 0.03 kg moisture per kilogram of dry air. Compute (i) the amount of water removed per 100 kg of wet material fed to the drier, (ii) the amount of dry air in fresh air per 100 kg of wet material, and (iii) the amount of dry air in recycled air per 100 kg of wet material.

Solution

Basis: 100 kg of wet material entering the drier

Figure 4.5 represents the data provided in the form of a block diagram.

(i) Feed to the drier contains 15% moisture and 85% moisture-free solids. The output from the drier contains 7% moisture and 93% moisture-free solids.

Moisture-free solids in 100 kg of wet material entering the drier = 85 kg

This forms 93% of the output from the drier.

Weight of material leaving the drier $= \dfrac{85 \times 100}{93} = 91.4$ kg

Water removed per100 kg of feed $= 100 - 91.4 = 8.6$ kg

(ii) Dry air entering the drier contains 0.03 kg moisture per kg of dry air. It picks up moisture from the wet solid and leaves the drier with a moisture content of 0.1 kg per kilogram of dry air.

1 kg dry air picks up 0.07 kg moisture.

Moisture to be picked up from 100 kg of wet material fed to drier = 8.6 kg

Dry air required for this $= 8.6 \div 0.07$

$= 122.86$ kg

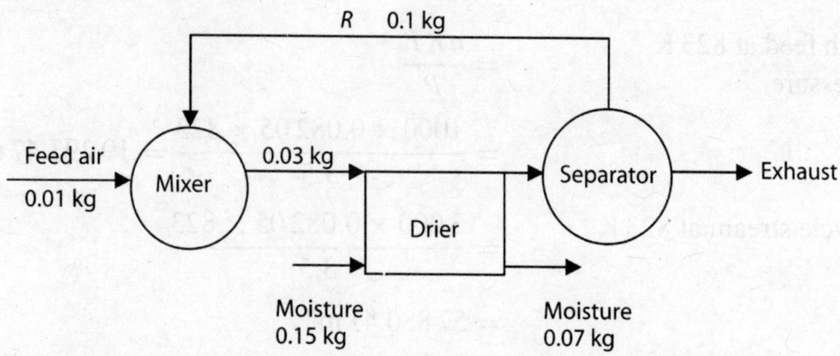

Fig. 4.5 Drying operation involving recycle

Air fed to the drier is obtained by mixing recycled air and fresh air.

Fresh air contains 0.01 kg moisture per kilogram of dry air; recycle air contains 0.1 kg moisture per kilogram of dry air; mixed air fed to the drier contains 0.03 kg moisture per kilogram of dry air.

Mixed air fed to the drier as calculated above is 122.86 kg.

Let dry air in the recycle air be R kg.

Then dry air in fresh air becomes $(122.86 - R)$ kg.

R can be calculated as given below using moisture balance over the mixer.

Moisture in mixed air = moisture in recycled air + moisture in fresh air

$122.86 \times 0.03 = (122.86 - R) \times 0.01 + 0.1R$

Dry air in recycle air $(R) = 27.33$ kg

Dry air in fresh air $= 95.53$ kg

4W Methanol is produced by the reaction of carbon monoxide with hydrogen according to the reaction

$$CO + 2H_2 \rightarrow CH_3OH$$

The product methanol is condensed and separated from unconverted gases, which are recycled. If 15% of carbon monoxide entering the reactor is converted to methanol, calculate (i) the volume of feed gas, and (ii) the volume of recycle gas required to produce 3 200 kg h^{-1} of methanol, if the gases enter the converter at 298 K and 1 bar pressure.

Solution

Basis: 3 200 kg h^{-1} production of methanol

Methanol to be produced $= 3\,200 \div 32 = 100$ kmol

Methanol is produced by the reaction of carbon monoxide with hydrogen as per the reaction

$$CO + 2H_2 \rightarrow CH_3OH$$

Each mole of methanol requires 3 mol of carbon monoxide and hydrogen.

Amount of reactants required for 100 kmol methanol $= 300$ kmol

But conversion in reactor is only 15%.

Reactants to be fed to the reactor for 100 kmol methanol $= 300 \div 0.15 = 2\,000$ kmol

Of this, 300 kmol is converted to methanol and the balance is recycled.

Amount of recycled reactants $= 2\,000 - 300 = 1\,700$ kmol

Volume of recycle at 298 K and 1 bar pressure $= \dfrac{nRT}{P}$

$$= \frac{1\,700 \times 0.082\,05 \times 298 \times 1.013}{1} = 42\,106.98\ \text{m}^3$$

(1 atm pressure = 1.013 bar. So 1 bar = 1 ÷ 1.013 atm)

For the process to continue, 2 000 kmol is required, which is obtained by mixing recycle with fresh feed.

So fresh feed required $= 300$ kmol

Volume of fresh feed at 298 K and 1 bar pressure $= \dfrac{300 \times 0.082\,05 \times 298 \times 1.013}{1}$

$$= 7\,430.63\ \text{m}^3$$

4X Formaldehyde is produced in a catalytic reactor by dehydrogenation of methanol; per pass conversion of methanol is 65%. Unreacted methanol is separated from the product stream and recycled to the reactor. Calculate the flow rate of fresh feed and recycle feed required for the production of 3 000 kg h^{-1} for formaldehyde. Compute the recycle ratio.

Solution

Basis: 3 000 kg h^{-1} of formaldehyde

Formaldehyde to be produced $= 3\,000 \div 30 = 100$ kmol

The reaction involved is

$$CH_3OH \rightarrow HCHO + H_2$$

Conversion per pass is only 65%.

Methanol required for 100 kmol formaldehyde $= 100 \div 0.65 = 153.85$ kmol

Feed to the reactor $= 153.85$ kmol

Methanol converted $= 153.85 \times 0.65 = 100$ kmol

Methanol recycled $= 53.85 \text{ kmol h}^{-1}$

To maintain continuous production, fresh methanol equivalent to that consumed in the reaction is to be fed.

Fresh methanol fed to the reactor $= 100 \text{ kmol h}^{-1}$

Recycle ratio is the ratio of recycle to fresh feed.

Recycle ratio $= 53.85 \div 100 = 0.5385$

4Y A plant produces material at the rate of 4 000 kg h^{-1} containing 75% water and 25% solids. The material is first granulated and then dried to a final moisture content of 16.67%. The feed to the granulator must have a water content of 50% only. For this, fresh feed is mixed with part of the final product leaving the drier. Air entering the drier contains 3% moisture, while air leaving the drier contains 6% moisture. Compute the recycle rate to the granulator and rate of hot air flow to the drier.

Solution

Basis: 1 hour operation

Figure 4.6 represents the data provided in the form of a block diagram.

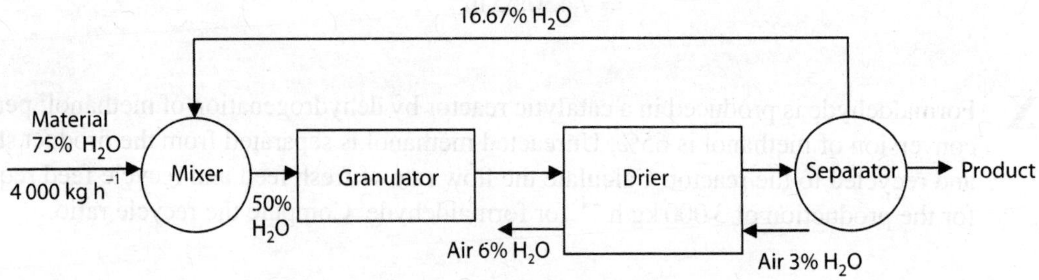

Fig. 4.6 Drying operation involving recycle

Feed to the granulator is prepared by mixing together fresh feed containing 75% moisture, and part of the final product containing 16.67% moisture, which is recycled. Moisture content of the mixed feed is 50%.

Fresh feed is 4 000 kg. Let the recycle be R kg.

Moisture balance over the mixer preparing the feed for the granulator

Moisture in fresh feed + moisture in recycle = moisture in the output

$$\frac{4\,000 \times 75}{100} + \frac{R \times 16.67}{100} = \frac{(4\,000 + R) \times 50}{100}$$

$$R = 3\,000\,\text{kg}$$

Feed to the granulator per hour $= 4\,000 + 3\,000 = 7\,000\,\text{kg}$

Moisture in material entering the drier $= 7\,000 \times 0.5 = 3\,500\,\text{kg}$

The balance, moisture-free solids, is 3 500 kg. This forms 83.33% (100%–16.67%) of the product leaving the drier.

Moisture in the material leaving the drier $= \dfrac{3\,500 \times 16.67}{83.33} = 700.17\,\text{kg}$

Moisture lost in the drier = moisture entering – moisture leaving

$= 3\,500 - 700.17 = 2\,799.83\,\text{kg}$

Air entering the drier contains 3% moisture. It picks up moisture in the drier and leaves with a moisture content of 6%.

97 kg of dry air in 100 kg air entering the drier becomes 94% in the air leaving the drier.

100 kg feed air becomes or $\dfrac{100 \times 97}{94}$ 103.19 kg outgoing air

Moisture pickup per 100 kg feed air $= 103.19 - 100 = 3.19\,\text{kg}$

Air to be fed to the drier per hour $= \dfrac{2\,799.83 \times 100}{3.19} = 87\,769\,\text{kg}$

4Z Air at 1 bar and 303 K containing 0.05 mol moisture per mole of dry air is to be cooled to 298 K containing 0.01 mol moisture per mole of dry air. The cooled air is mixed with part of the original air which bypasses the cooler. The resulting mixture is heated back to 303 K and it has 0.02 mol moisture per mole of dry air. For every 200 m^3h^{-1} of leaving air, determine the volume of entering air and the percentage bypass.

Solution

Basis: 200 m^3 of leaving air

Figure 4.7 represents the data provided in the form of a block diagram.

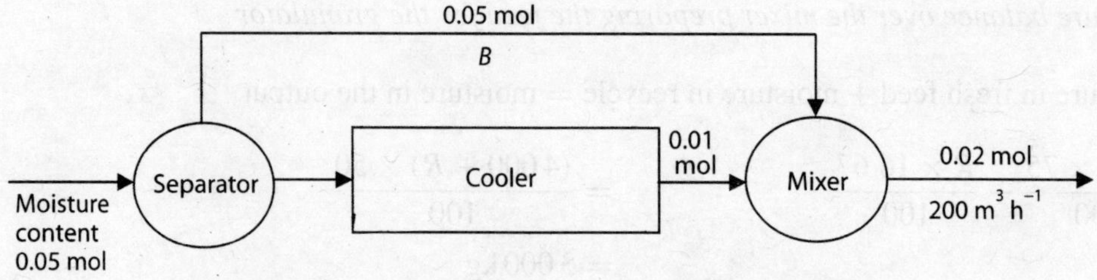

Fig. 4.7 Air conditioning operation involving bypass

$$\text{Number of kilomoles of leaving air } (n) = \frac{PV}{RT} = \frac{200 \times 1}{1.013 \times 0.082\,05 \times 303} = 7.941\,\text{kmol}$$

Leaving air contains 0.02 mol moisture per mole of dry air.

In other words 1.02 mol of leaving air contains 1 mol dry air.

Dry air in the leaving air $= 7.941 \div 1.02 = 7.786\,\text{kmol}$

Dry air required for the process $= 7.786\,\text{kmol}$

Of this, let B kmol bypass the cooler.

Dry air entering the cooler $= (7.786 - B)\,\text{kmol}$

The unknown, B, can be found by a moisture balance around the mixer.

Moisture content of the product from the cooler is 0.01 mol per mole of dry air.

Moisture content of the air bypassing the cooler is 0.05 mol per mole of dry air.

Moisture content of the product from the mixer is 0.02 mol per mole of dry air.

Moisture entering the mixer = moisture leaving the mixer

Moisture content of the product from the cooler + moisture content of the air bypassing the cooler = moisture content of the product from the mixer

$$0.05B + 0.01(7.786 - B) = 7.786 \times 0.02$$

$$\therefore \quad B = 1.946\,5\,\text{kmol}$$

$$\text{Percentage of bypass} = \frac{1.946\,5 \times 100}{7.786} = 25\%$$

Moisture content of air entering the process is 0.05 mol per mole dry air.

Entering wet air $= 7.786 \times 1.05 = 8.175\,3\,\text{kmol}$

$$\text{Volume of this air at 303 K and 1 bar} = \frac{8.175\,3 \times 0.082\,05 \times 303 \times 1.013}{1}$$

$$= 205.89\,\text{m}^3$$

4AA Brackish water is desalinated by evaporation for irrigation purposes. If the feed water has 1000 ppm of salt, and if the product has to contain $5 \times 10^{-3}\%$ salts, what fraction of feed water should bypass the evaporator?

Solution

Basis: 1 kg of feed

Figure 4.8 represents the data provided in the form of a block diagram.

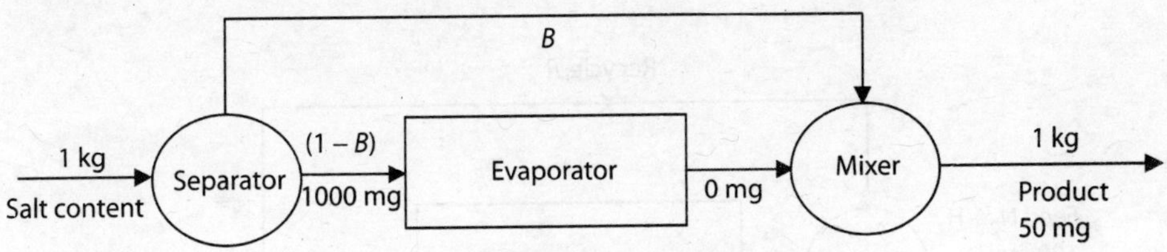

Fig. 4.8 Desalination involving bypass

A certain fraction of the feed is fed to an evaporator and the remaining feed bypasses the evaporator. The salt content of the output from the evaporator is zero. Water for irrigation is obtained by mixing together the output from the evaporator and the feed bypassing the evaporator.

Let B kg of feed bypass the evaporator. The unknown, B, can be found by a moisture balance around the mixer.

Salt content of feed = 1000 mg (ppm is also mg per kg)

$$\text{Salt content of product} = \frac{1000 \times 5 \times 100}{1000} = 50 \text{ ppm} = 50 \text{ mg}$$

Salt balance over the mixer

Salt content of the product from the evaporator + salt content of water bypassing the evaporator = salt content of the product from the mixer

$$0(1 - B) + 1000B = 50$$

$$\therefore \quad B = 0.05$$

$$\text{Fraction bypassed} = \frac{1}{20}$$

4BB In the operation of a synthetic ammonia plant, a 1:3 nitrogen to hydrogen mixture is fed to the converter resulting in 25% conversion to ammonia. Ammonia formed is separated by condensation and the unconverted gases are recycled to the reactor. The initial nitrogen and hydrogen mixture contains 0.2 part of argon to 100 parts of nitrogen and hydrogen mixture. The toleration limit of argon entering the converter is 5 parts of argon to 100 parts of nitrogen and hydrogen by volume. Estimate the amount of recycle that must be continuously purged.

Solution

Basis: 100 mol of fresh synthesis gas fed

Figure 4.9 represents the data provided in the form of a block diagram.

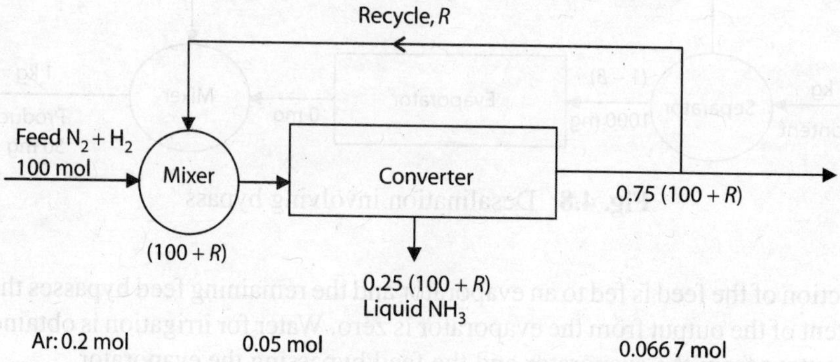

Fig. 4.9 Ammonia synthesis involving recycle and purge

Fresh synthesis gas is mixed with recycle and fed to the converter. Argon content of the synthesis gas mixture entering the converter is 0.05 mol per mole of synthesis gas.

25% of the synthesis gas entering the converter is converted to ammonia, which is removed by liquefaction and the remaining gases are recycled. Argon initially present in (100 + R) mol will now be present in 75% of (100 + R) mol.

$$\text{Argon content of the recycle} = 0.05 \times \frac{100}{75} = 0.0667 \text{ mol per mol of gas recycled.}$$

To maintain the argon content within the toleration limit of 0.05 mole per mole of gas even after mixing with fresh feed, argon entering by way of fresh feed has to be removed from the system by purging.

Argon accompanying 100 mol fresh synthesis gas = 0.2 mol

To remove 0.2 mol argon from the system,

$$\text{Amount of recycle gas to be purged} = \frac{0.2}{0.0667} = 3.0$$

For every 100 mol of fresh feed, 3 mol of recycle has to be purged.

4CC Sea water is desalinated in a reverse osmosis cell to produce potable water containing 500 ppm sodium chloride. The concentrated effluent from the RO cell contains 6% sodium chloride. To conserve water, part of it is recycled. Fresh seawater containing 3% sodium chloride and the recycle stream are mixed and the mixed feed containing 4% sodium chloride is fed to the RO cell. The concentrated effluent not recycled is purged out. Based on 1000 kg per hour of sea water fed to the RO cell, calculate (i) rate of waste brine purged out, (ii) rate of production of potable water, and (iii) fraction of waste brine recycled.

Solution

Basis 1 hour operation

Figure 4.10 represents the data provided in the form of a block diagram.

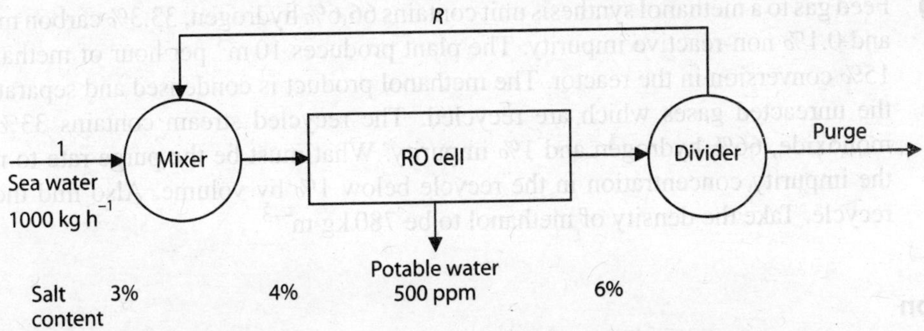

Fig. 4.10 Reverse osmosis involving recycle and purge

Feed to the RO cell is prepared by mixing seawater and recycled waste brine.

Salt content of seawater is 3%, that of waste brine is 6%, while that of the mixed feed to the RO cell is 4%.

Feed rate of seawater is 1000 kg per hour.

Let the flow rate of waste brine recycled be R kg per hour.

Feed rate to the RO cell is $(1000 + R)$ kg per hour.

The unknown, R, can be found by a salt balance over the mixer.

Salt in sea water + salt in recycle = salt in mixed feed

$$\left(\frac{3 \times 1000}{100}\right) + \left(\frac{6 \times R}{100}\right) = \frac{4 \times (1000 + R)}{100}$$

$$\therefore \quad R = 500 \text{ kg per hour}$$

Feed rate to the RO cell is 1 500 kg per hour. The output of the RO cell is potable water with 500 ppm salt content and waste brine with a salt content of 6%.

Let the rate of production of potable water be D kg per hour.

The unknown, D, can be found by a salt balance over the RO cell.

Salt in mixed feed = salt in potable water + salt in waste brine

$$\frac{4 \times 1\,500}{100} = \frac{500 \times D}{1\,000\,000} + \frac{6 \times (1\,500 - D)}{100}$$

$$D = 504.2 \text{ kg per hour}$$

Waste brine $= 1\,500 - 504.2 = 995.8$ kg per hour

Of this 500 kg per hour is recycled.

$$\text{Fraction of waste brine recycled} = \frac{500}{995.8} = 0.502$$

4DD Feed gas to a methanol synthesis unit contains 66.6% hydrogen, 33.3% carbon monoxide and 0.1% non-reactive impurity. The plant produces 10 m^3 per hour of methanol with 15% conversion in the reactor. The methanol product is condensed and separated from the unreacted gases which are recycled. The recycled stream contains 33% carbon monoxide, 66% hydrogen and 1% impurity. What must be the purge rate to maintain the impurity concentration in the recycle below 1% by volume. Also find the rate of recycle. Take the density of methanol to be 780 kg m^{-3}.

Solution

Basis: 1 hour of operation

Figure 4.11 represents the data provided in the form of a block diagram.

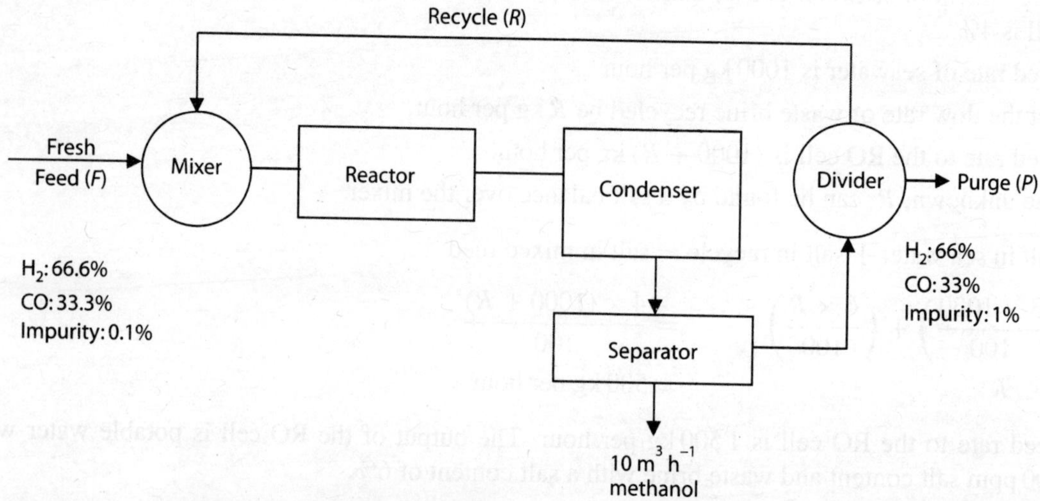

Fig. 4.11 Methanol synthesis involving recycle and purge

$$\text{Methanol produced} = 780 \times 10\,\text{kg} = \frac{780 \times 10}{32} = 243.75\,\text{kmol}$$

Let the feed rate, recycle rate and purge rate be F, R and P respectively. The unknowns can be found using appropriate material balance equations. The feed contains 33.3% carbon monoxide and the recycle stream from the reactor contains 33% carbon monoxide.

Overall carbon balance

$$\begin{aligned} 0.333F &= \text{Methanol product} + 0.33(R+P) \\ &= 243.75 + 0.33(R+P) \end{aligned} \tag{1}$$

When steady state is attained, the quantity of impurity in the system is constant. The non-reactive impurity cannot leave along with methanol product. All the impurity from the fresh feed must leave in the purge.

Overall impurity balance

$$0.001F = 0.01P \tag{2}$$

From equations (1) and (2)

$$F = 812.5\,\text{kmol h}^{-1}$$
$$P = 81.25\,\text{kmol h}^{-1}$$

So purge is equal to 10% of fresh feed. The reaction occurring in the converter is

$$CO + 2H_2 \rightarrow CH_3OH$$

1 mol of methanol requires 3 mol of $CO + H_2$ mixture. The reaction goes to only 15% completion. Unreacted reactants are separated from the product and recycled to the reactor.

$$\text{Mixed feed required for 243.75 kmole methanol} = \frac{243.75 \times 3 \times 100}{15}$$
$$= 4\,875.0\,\text{kmol}$$

Fresh feed rate per hour $= 812.5$ kmol

Recycle feed rate per hour $= 4\,062.5$ kmol

Problems for Practice

4.1 500 kg of 90% pure sulphur is burnt with 25% excess air over and above that required to convert all sulphur to sulphur dioxide. Only 95% of sulphur is burnt. Of the sulphur burnt, 5% goes to sulphur trioxide and the balance to sulphuric dioxide. Calculate the composition of the gases leaving the burner. If the gases leave at 673 K and 750 mm Hg pressure, compute the volume of the gases.

4.2 Acetylene is produced by treating calcium carbide with water. Calculate the number of hours of service that can be derived from 10 kg of carbide in an acetylene lamp burning 60 L of gas per hour at a temperature of 293 K and a pressure of 74 cm mercury.

4.3 Nitrobenzene is produced by nitrating benzene with mixed acid containing 39% nitric acid, 53% sulphuric acid and balance water. The reaction is 97% complete. 1000 kg of benzene and 2 070 kg of mixed acid are used. Calculate (i) the quantity of nitrobenzene produced, and (ii) the quantity and composition of the spent acid mixture.

4.4 Chlorine is manufactured by the oxidation of hydrogen chloride gas. If 50% excess air is used for the reaction, calculate (i) the weight of air used per kg of acid, (ii) the composition of the gases entering the reactor, and (iii) the composition of the gases leaving the reactor, assuming the reaction to be 60% complete.

4.5 Antimony is produced by heating powdered antimony trisulphide with iron, and removing molten antimony from the reaction vessel. If 1.8 kg of antimony trisulphide and 1.3 kg of iron are heated to give 0.9 kg of antimony, identify the limiting reactant and excess reactant. Calculate the percentage excess reactant, degree of completion and percentage conversion.

4.6 Pure sulphur is burnt in a burner at the rate of 500 kg per hour. Air is supplied at 303 K and 735 mm Hg. The gases from the burner contain 16.5% sulphur dioxide, 3% oxygen and the rest nitrogen. The gases leave the burner at 1 073 K and 760 mm Hg. Calculate (i) the fraction of sulphur burnt to sulphur trioxide, (ii) the percentage of excess air used, and (iii) the volume of dry air used in $m^3\,h^{-1}$.

4.7 Carbon tetrachloride is made from carbon disulphide and chlorine. The gases are found to contain 33.3% of carbon tetrachloride, 33.3% of sulphur monochloride, 1.4% of carbon disulphide and the rest chlorine. Calculate (i) the percentage excess reactant, (ii) the percentage conversion, and (iii) the weight of carbon tetrachloride produced per kilogram of chlorine.

4.8 Pure sulphur is burnt in a furnace. Even when 20% excess dry air is passed, only 30% of sulphur is burnt to sulphur trioxide (the desired reaction) and the rest to sulphur dioxide. What is the analysis of the resulting gases? The gases from the burner are passed through a catalytic converter where sulphur dioxide is oxidised to sulphur trioxide without the addition of any more air. If the gases leaving the converter contain 4.3% sulphur dioxide, calculate the composition of exit gases from the converter.

4.9 In the Le Blanc process the following reaction takes place:

$$2\,NaCl + H_2SO_4 \rightarrow NaHSO_4 + HCl + NaCl$$

The aqueous acid used contains 80% sulphuric acid. It is supplied in 5% excess of that theoretically required. The reaction goes to completion. 1000 kg of salt are charged per batch. (i) Calculate the weight of acid supplied. (ii) If 90% of hydrogen chloride formed and 25% of water present are removed, compute their amounts. (iii) Find the composition of the remaining salt mixture.

4.10 A rich copper ore on analysis shows 10% cupric sulphide, 30% ferrous sulphide and balance inerts. By crushing and floatation process, two thirds of the inerts are eliminated. The resulting concentrate on analysis shows 16% cupric sulphide, 48% ferrous sulphide and balance inerts. The ore is roasted with carbon. In the reaction 5% of copper is lost along with unleached inerts. Find the weight of copper obtainable from 1 tonne of ore.

4.11 Phosphorous is prepared by heating a thoroughly mixed mass of calcium phosphate, sand and charcoal in an electric furnace. In the charge fed to the furnace, the amount of silica used is 10% in excess, and charcoal is 40% in excess of that required to combine with phosphorous pentoxide liberated to form carbon monoxide. For 100 kg of charge, calculate (i) the weight of the constituents and composition of the charge, and (ii) the weight of phosphorous obtained. The reaction of calcium phosphate with silica is 90% complete and reaction of phosphorous pentoxide with carbon is 70% complete.

4.12 Pyrites containing 35% iron, 40% sulphur and the rest inerts are roasted. The gases leaving the roaster at 773 K have the following composition: sulphur dioxide 9%, oxygen 8.6% and the rest nitrogen. For 100 kg pyrites roasted, calculate (i) the volume of air supplied at standard conditions, (ii) the percentage of excess air, and (iii) the volume of the gases leaving in cubic metres.

4.13 Benzene reacts with nitric acid to produce nitrobenzene and water. Nitrobenzene formed may undergo further nitration to form dinitrobenzene. The percentage conversion of benzene is 90% and the acid used is 65% in excess over the theoretical requirement for the formation of nitrobenzene. If the molar ratio of nitrobenzene to dinitrobenzene in the product stream is 17:1, calculate the quantities of benzene and nitric acid required for the production of 1000 kg per hour of nitrobenzene.

4.14 Analysis of a gas entering a secondary converter in a contact sulphuric acid plant gave the result: 4% sulphur dioxide, 13% oxygen and the rest nitrogen. The gases leaving the converter have 0.45% sulphur dioxide on sulphur trioxide-free basis. Calculate the percentage of sulphur dioxide converted.

4.15 Formaldehyde is produced by the catalytic oxidation of methanol using an excess of air according to the reaction:

$$CH_3OH + (O) \rightarrow HCHO + H_2O$$

Due to improper control the following side reaction also takes place.

$$HCHO + (O) \rightarrow HCOOH$$

In a typical run, the product gases from the reactor have the following composition by volume: methanol 8.6%, formaldehyde 3.1%, formic acid 0.6%, water vapour 3.7%, oxygen 16% and nitrogen 68%. Calculate (i) the percentage conversion of methanol to formaldehyde, (ii) the percentage of methanol lost in the side reaction, and (iii) the molar ratio of methanol to air in the feed.

4.16 Gaseous benzene reacts with hydrogen in the presence of nickel catalyst to give cyclohexane. 30% excess hydrogen is used, the conversion is 50% and the yield is 90%. Calculate the requirement of benzene and hydrogen gas for the production of 1000 kg per hour of cyclohexane.

4.17 Orthoxylene on oxidation gives phthalic anhydride. 20% excess of air is used. The conversion is 60% and the yield is 75%. Calculate the requirement of orthoxylene and air in the production of 1000 kg per hour of phthalic anhydride.

4.18 Benzaldehyde is manufactured by the catalytic oxidation of toluene with dry air. To get a high yield, 100% excess air is used. 0.5% of toluene charged burns to carbon dioxide and water by an unavoidable side reaction. If the overall conversion of toluene is 13%, calculate the composition of the gas stream leaving the reactor.

4.19 An ammonia converter is fed with a 1:3 mixture of nitrogen and hydrogen. The conversion of ammonia is 27% based on nitrogen. The ammonia formed is separated by liquefaction and the unconverted reactants are recycled back to the converter. Find (i) the volume of fresh feed, and (ii) the volume of recycle to produce 7 200 kg per hour of ammonia. Fresh feed and recycle are at 823 K and 2.5 atm pressure.

4.20 A 1:2 mixture of carbon monoxide and hydrogen is fed to a converter in which 25% conversion to methanol takes place based on carbon monoxide. The product formed is separated completely and unconverted reactants are recycled back to the converter. Find (i) the volume of fresh feed, and (ii) the volume of recycle to produce 6 400 kg h^{-1} of methanol. Fresh feed and recycled stream are at 318 K and 3.5 atm.

4.21 It is desired to make a sodium hydroxide solution of composition 30wt%. Due to high heat of dissolution of caustic in water, the above is prepared in a two-step process. First, all sodium hydroxide is dissolved in the correct quantity of water to produce 50wt% caustic. After complete dissolution and cooling, this solution is diluted with water to produce the desired solution. Calculate the weight fraction of water that has bypassed the dissolution stage.

4.22 In the production of ammonia from hydrogen and nitrogen, the conversion based on raw material is limited to 15%. Ammonia produced is condensed from the reactor product stream and unreacted material is recycled. If the feed contains 0.2% argon, calculate the purge rate required to hold the amount of argon in the recycle stream below 5%.

4.23 Aniline is produced by the hydrogenation of nitrobenzene. A small amount of cyclohexyl amine is also produced as a by-product. The reactions are

$$C_6H_5NO_2 + 3\,H_2 \rightarrow C_6H_5NH_2 + 2\,H_2O$$
$$C_6H_5NO_2 + 6\,H_2 \rightarrow C_6H_{11}NH_2 + 2\,H_2O$$

Nitrobenzene along with three times the stoichiometric quantity of hydrogen is fed to the reactor. Conversion of nitrobenzene is 96% and the yield of aniline is 95%. Unreacted hydrogen is separated from the reactor products and recycled to the reactor. A purge is taken from the recycle stream to maintain the inerts in the recycle stream below 5%. Fresh hydrogen feed is 99.5% pure, the remainder being inerts. If all percentages are by mole, for a feed of 100 kmol per hour of nitrobenzene, calculate (i) the fresh hydrogen feed, (ii) the purge rate required, and (iii) the composition of reactor outlet stream.

4.24 60wt% nitric acid produced in a process has to be concentrated to 99wt% nitric acid. Due to the formation of an azeotrope, nitric acid cannot be concentrated beyond 68wt%. To overcome this, 72wt% magnesium nitrate solution is used as an entrainer. The feed is mixed with the entrainer and distilled. The distillate is the required product. 55wt% magnesium nitrate solution obtained as residue is concentrated by evaporation to recover 72wt% magnesium nitrate solution for reuse. For the production of 1000 kg h^{-1} of 99wt% nitric acid, compute the entrainer requirement and the amount of water evaporated.

4.25 A petroleum fraction of composition 80% n-pentane and 20% isopentane is produced in a petroleum refinery. To manufacture gasoline from this, isopentane content of this stream has to be reduced to 10% using an isopentane remover. What fraction of the feed can bypass the isopentane remover?

4.26 An airconditioning plant has to supply 1000 m^3 min^{-1} of air at 298 K with a water vapour content of 0.02 mol per mole of dry air to an auditorium. The plant draws atmospheric air at a temperature of 308 K with a water vapour content of 0.05 mol per mole of dry air, cools part of it to a water vapour content of 0.01 mol per mole of dry air and mixes it with uncooled air. The resulting mixture is then reheated to 298 K and supplied to the auditorium. Calculate the volume of air entering and the percentage of air entering that bypasses the cooler.

5 Material Balance in Unit Operations

Almost all the activity of industrial chemical production can be broken down into a series of steps called *unit operations*, the principles of which are independent of the material being processed. Some of these unit operations involve chemical reactions whereas others are purely physical in nature. The process used in the manufacture of common salt consists of the following sequence of unit operations: transportation of solids and liquids, transfer of heat, evaporation, crystallisation, drying and screening. No chemical reaction appears in these steps. On the other hand, cracking of petroleum is a typical chemical reaction conducted on an enormous scale. Here the unit operations—transportation of fluids and solids, and distillation and various mechanical separations—are vital and the cracking reaction cannot be carried out without them.

GAS ABSORPTION

The selective removal of one of the components of a mixture of gases by contact with a suitable liquid solvent is known as *absorption*. The operation is also known as *scrubbing*. A major application of absorption technology is the removal of carbon dioxide and hydrogen sulphide from natural gas or synthesis gas by absoption in aqueous solutions of mono-, di- or triethanol amines or alkaline salts. Another example is the washing of ammonia from a mixture of ammonia and air by water. Absorption may be a mere physical activity or accompanied by a chemical reaction. A schematic representation of a typical gas absorption operation is given in Figure 5.1.

In absorption, the feed is a gaseous mixture and is introduced at the bottom of a packed tower, which often acts as an absorption column. The liquid solvent is fed from the top. As the solvent trickles down the column, it selectively absorbs one of the components of the gaseous mixture and leaves with it at the lower end of the column. The unabsorbed components of the gaseous mixture leave the column from the top.

If the composition of the gaseous feed, feed rate and the percentage of absorption in the tower are known, the rate and composition of the gaseous output from the absorption column can be calculated. If the liquid feed rate is also known, the rate and composition of the liquid effluent can also be computed.

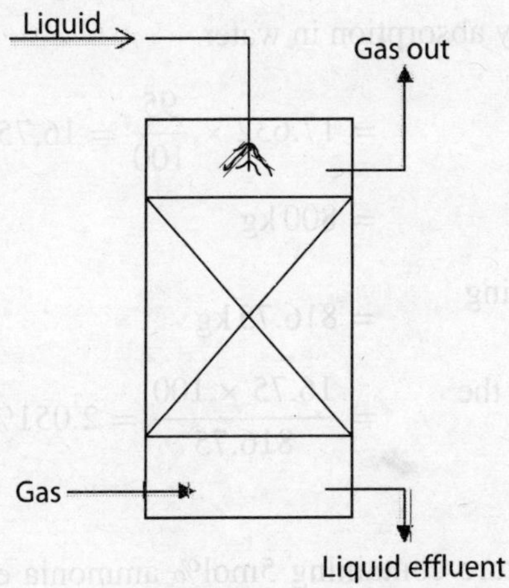

Fig. 5.1 Schematic representation of a typical gas absorption operation

ILLUSTRATIONS

5A It is desired to absorb 95% acetone from a feed mixture of acetone and air containing 2mol% acetone, using pure water. Pure water at the rate of 800 kg h^{-1} is introduced at the top of the tower and the gas flow rate at the bottom of the tower is 450 kg h^{-1}. Find the composition of acetone in the liquid effluent flowing out of the tower.

Solution

Basis: 1 hour of operation

The composition of the gas is given in mole per cent, and the gas flow rate is given in terms of mass. Gas flow rate has to be converted to molar basis to find its acetone content.

Gas fed to the tower contains 2mol% acetone and 98mol% air.

Average molecular weight of air is 29 and the molecular weight of acetone is 58.

Average molecular weight of the gas fed to the tower $= 58 \times 0.02 + 29 \times 0.98 = 29.58$

Number of kilomoles of the gas is found by dividing the mass by the average molecular weight.

Amount of gas entering $= 450 \div 29.58 = 15.213$ kmol

Amount of acetone in the entering gas $= \dfrac{15.213 \times 2}{100} = 0.304$ kmol

$= 0.304 \times 58 = 17.632$ kg

95% of acetone is removed by absorption in water.

Weight of acetone in water $= 17.632 \times \dfrac{95}{100} = 16.75\,\text{kg}$

Water in the effluent from the absorber $= 800\,\text{kg}$

Weight of liquid effluent leaving the absorber $= 816.75\,\text{kg}$

Weight per cent of acetone in the liquid effluent flowing out $= \dfrac{16.75 \times 100}{816.75} = 2.051\%$

5B An air–ammonia mixture containing 5mol% ammonia enters a packed tower at the rate of 500 kmol h^{-1}. It is desired to recover 90% of ammonia using 700 kmol h^{-1} of pure water. Compute the composition of ammonia in the liquid effluent and gas effluent from the absorption tower.

Solution

Basis: 1 hour of operation

Amount of gas entering the absorber = 500 kmol

Entering gas is an air–ammonia mixture. It contains 5mol% ammonia.

Amount of ammonia in the gas entering the absorber $= \dfrac{500 \times 5}{100} = 25\,\text{kmol}$

Air in the gas entering the absorber $= 475\,\text{kmol}$

90% ammonia is absorbed by water.

Ammonia absorbed by water $= \dfrac{25 \times 90}{100} = 22.5\,\text{kmol}$

Ammonia in the outgoing air = ammonia entering − ammonia absorbed

$= 25 - 22.5 = 2.5\,\text{kmol}$

Water absorbs only ammonia. Amount of air in the outgoing gas is the same as in the entering gas.

Amount of outgoing gas = air in entering gas + unabsorbed ammonia

$= 475 + 2.5 = 477.5\,\text{kmol}$

Mole per cent of ammonia in the gaseous effluent from the tower $= \dfrac{2.5 \times 100}{477.5} = 0.524\%$

The valid assumption for expressing the composition of gaseous effluent is mole per cent.

Water fed to the tower $= 700\,\text{kmol} = 700 \times 18 = 12\,600\,\text{kg}$

Weight of ammonia absorbed by water $= 22.5\,\text{kmol} = 22.5 \times 17 = 382.5\,\text{kg}$

Molecular weights of water and ammonia are 18 and 17 respectively.

Weight of the liquid effluent $= 12\,982.5\,\text{kg}$

Weight per cent of ammonia in the liquid effluent from the tower $= \dfrac{382.5 \times 100}{12\,982.5} = 2.946\%$

The valid assumption for expressing the composition of the liquid effluent is weight per cent.

5C Benzene–air mixture is to be scrubbed in a counter-current absorption tower using a non-volatile hydrocarbon oil as a solvent. Inlet gas contains 10% benzene and the entering gas flow rate is $1\,500\,\text{m}^3\,\text{h}^{-1}$. It is proposed to remove 80% of the benzene in the entering gas. The average molecular weight of the oil is 200. Find the required oil flow rate in kilograms per hour and the composition of the outgoing gases. The mole fraction of benzene in the liquid outlet from the tower has to be 0.3.

Solution

Basis: 1 hour of operation

The gas input to the tower is on a volume basis. The valid assumption for expressing the composition of the gaseous mixture is mole per cent . The gas composition is in mole per cent . Gas feed rate has to be converted to molar basis, using the relation,

Amount of gas entering (n) $= \dfrac{PV}{RT} = \dfrac{1\,500 \times 1}{0.082\,05 \times 303} = 60.34\,\text{kmol}$

The gas entering contains 10mol% benzene.

Amount of benzene in gas entering $= \dfrac{60.34 \times 10}{100} = 6.034\,\text{kmol}$

80% of benzene in the gas entering is recovered by absorption in hydrocarbon oil.

Amount of benzene absorbed in hydrocarbon oil $= \dfrac{6.034 \times 80}{100} = 4.827\,\text{kmol}$

Mole fraction of benzene in the liquid outlet from the tower has to be 0.3.

$\therefore$ Mole fraction of hydrocarbon oil in the liquid outlet from the tower is 0.7.

Amount of hydrocarbon oil in the liquid output $= \dfrac{4.827 \times 0.7}{0.3} = 11.263\,\text{kmol}$

The average molecular weight of the oil is 200.

Rate of oil added to the tower $= 11.263 \times 200 = 2\,252.6\,\text{kg h}^{-1}$

Amount of benzene-free gas entering the tower $=$ amount of entering gas − amount of benzene in the entering gases

$$= 60.340 - 6.034 = 54.306\,\text{kmol}$$

Amount of benzene in the outgoing gas $=$ amount of benzene in the input − amount of benzene absorbed

$$= 6.034 - 4.827 = 1.207\,\text{kmol}$$

Amount of outgoing gases $=$ amount of benzene-free gas in the gas input + amount of unabsorbed benzene

$$= 54.306 + 1.207 = 55.513\,\text{kmol}$$

Composition of benzene in the outgoing gases $= \dfrac{1.207 \times 100}{55.513} = 2.174\,\text{mol\%}$

DISTILLATION

Distillation is a unit operation where a liquid–liquid mixture is separated, based on the difference in the boiling point of the constituents. The separation of a mixture of alcohol and water into its components, liquid air into nitrogen and oxygen, and crude petroleum into petrol, diesel, kerosene and other components are examples of distillation. A schematic representation of a typical distillation operation is given in Figure 5.2.

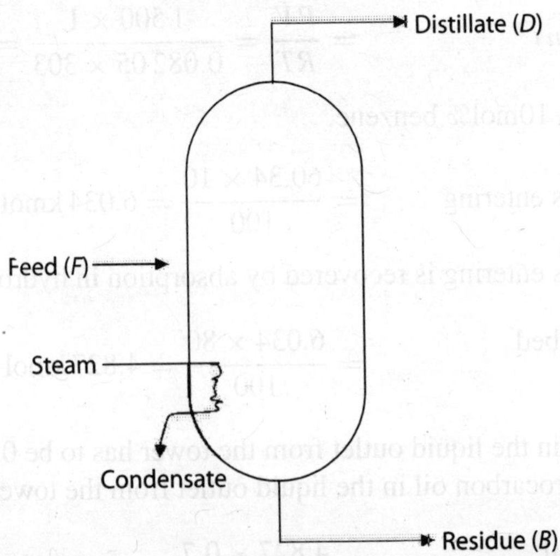

Fig. 5.2 Schematic representation of a typical distillation operation

The feed of a binary mixture of miscible and volatile components when heated in a distillation column splits into an overhead stream known as *distillate*, which is richer in the more volatile component and a bottom stream known as *residue*, which is richer in the less volatile component. In distillation problems, the composition of the feed and the feed rate are generally known. The desired purity levels of the distillate and residue dictate the composition and rate of these outputs.

ILLUSTRATIONS

5D A continuous fractioning column is to be designed to separate 10 000 kg h^{-1} of a mixture of 50mol% benzene and 50mol% toluene to give an overhead product containing 98mol% benzene and a bottom product of 97mol% toluene. Compute the overhead and bottom products obtained per hour.

Solution

Basis: 1 hour operation

The feed rate is given on weight basis. The composition of the product streams are given on molar basis. The feed rate also has to be converted to molar basis, so as to solve the problem on a consistent basis. Weight of the feed divided by the average molecular weight gives the number of kilomoles of the feed.

The feed contains 50mol% benzene and 50mol% toluene.

Average molecular weight of the feed $= 0.5 \times 78 + 0.5 \times 92 = 85$

Amount of feed (F) $= 10\,000 \div 85 = 117.65$ kmol

Let the number of kilomoles of distillate and residue be D and B.

We have two unknowns and to get their values, two equations are required. The required equations can be framed based on overall balance and benzene balance.

Mole fraction of benzene in feed (x_F) $= 0.50$

Mole fraction of benzene in distillate (x_D) $= 0.98$

Mole fraction of benzene in residue (x_B) $= 0.03$

Overall balance

$F = D + B$

$117.65 = D + B$ (1)

Benzene balance

$$F x_F = D x_D + B x_B$$

$$117.65 \times 0.50 = 0.98\,D + 0.03\,B \qquad (2)$$

We get $B = 59.45$ k mol h^{-1}; $D = 58.2$ kmol h^{-1}.

As the overhead product and bottom product are liquids, they have to be reported in terms of weight, which is the valid assumption for liquids. Average molecular weight multiplied by the number of moles gives the weight.

Bottom product contains 3mol% benzene and 97mol% toluene.

Average molecular weight of B $= 0.03 \times 78 + 0.97 \times 92 = 91.58$

Bottom product obtained $= 59.45 \times 91.58$

$= 5\,444.43$ kg h^{-1}

Top product contains 98mol% benzene and 2mol% toluene.

Average molecular weight of D $= 0.98 \times 78 + 0.02 \times 92 = 78.28$

Overhead product obtained $= 78.28 \times 58.2$

$= 4\,555.58$ kg h^{-1}

5E A mixture containing 20mol% butane, 35mol% pentane and the rest hexane is to be separated by fractional distillation into a distillate fraction with a composition of 95mol% butane, 4mol% pentane and the rest hexane and a bottom product. The distillate must contain 90% of the butane charged in the feed. Calculate the composition of the bottom product.

Solution

Basis: 100 mol of feed

Amount of butane in feed $= 20$ mol

Amount of pentane in feed $= 35$ mol

Amount of hexane in feed $= 45$ mol

It is given that the distillate must contain 90% of butane in the feed.

Butane in the distillate $= 20 \times 0.9 = 18$ mol

Composition of the distillate is 95mol% butane, 4mol% pentane and 1mol% hexane.

Pentane in distillate $= 4 \times \dfrac{18}{95} = 0.76$ mol

Hexane in distillate $= 1 \times \frac{18}{95} = 0.19\,\text{mol}$

Butane in bottom product = butane in feed – butane in distillate

$= 20 - 18 = 2\,\text{mol}$

Similarly, pentane in bottom product $= 35 - 0.76 = 34.24\,\text{mol}$

Hexane in bottom product $= 45 - 0.19 = 44.81\,\text{mol}$

Composition of the bottom product is calculated as given in Table 5.1.

Table 5.1 Composition of the bottom product

Constituent	Moles (mol)	Mole per cent
Butane	2.00	2.47
Pentane	34.24	42.25
Hexane	44.81	55.28
	81.05	

5F A feed mixture containing 40% benzene, 30% toluene and 30% xylene is fed to a distillation column. The distillate contains 99.5% benzene and 0.5% toluene. The residue from this unit is fed to another distillation column from which a distillate of composition 97% toluene, 2% benzene and 1% xylene and a residue of composition 5% toluene and 95% xylene are obtained. Compute the composition of the three final outputs from the system. Also find the composition of the intermediate stream.

Solution

Basis: 100 kg of feed

The system is in liquid phase. So the valid assumption for the composition is weight per cent The feed consists of 40 kg benzene and 30 kg each of toluene and xylene. The process is represented diagrammatically in Figure 5.3.

Feed F fed to the first distillation column is split into a distillate D_1 and residue B_1. Residue B_1 is fed to the second distillation column and is split into distillate D_2 and residue B_2. We have three unknowns and to get their values, three equations are required.

The required equations can be framed based on overall balance, benzene balance and toluene balance.

Overall balance

$$F = D_1 + D_2 + B_2 = 100 \quad (1)$$

99.5wt% of D_1 and 2wt% of D_2 are benzene.

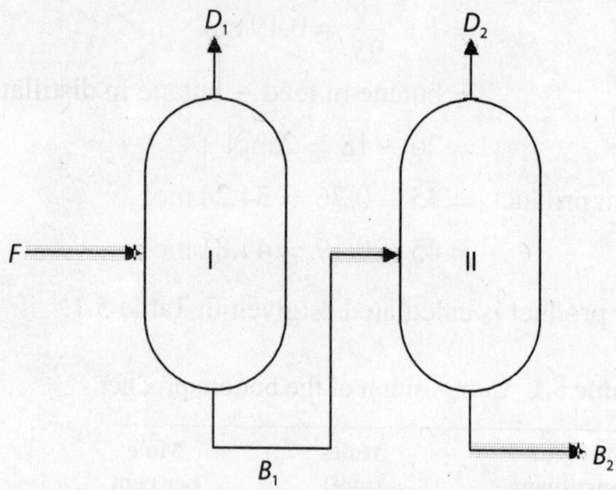

Fig. 5.3 Multicomponent distillation

Benzene balance

$$40 = 0.995D_1 + 0.02D_2 \qquad (2)$$

0.5wt% of D_1, 97wt% of D_2 and 5wt% of B_2 are toluene.

Toluene balance

$$30 = 0.005D_1 + 0.97D_2 + 0.05B_2 \qquad (3)$$

Solving equations (1), (2) and (3)

$$D_1 = 39.6\,\text{kg}$$
$$D_2 = 29.2\,\text{kg}$$
$$B_2 = 31.2\,\text{kg}$$

Weight of intermediate stream, $B_1 = F - D_1 = 100 - 39.6 = 60.4\,\text{kg}$

D_1 contains 99.5wt% benzene and the balance is toluene

Benzene in $D_1 = 39.6 \times 0.995 = 39.4\,\text{kg}$

Toluene in $D_1 = 39.6 - 39.4 = 0.2\,\text{kg}$

Toluene in B_1 = toluene in feed − toluene in D_1

$= 30 - 0.2 = 29.8\,\text{kg}$

Benzene in B_1 = benzene in feed − benzene in D_1

$= 40 - 39.4 = 0.6\,\text{kg}$

Xylene in B_1 = xylene in feed − xylene in D_1

$= 30 - 0 = 30.0\,\text{kg}$

Composition of the intermediate stream is calculated as given in Table 5.2.

Table 5.2 Composition of the intermediate stream

Constituent	Weight (kg)	Weight per cent
Benzene	0.6	0.99
Toluene	29.8	49.34
Xylene	30.0	49.67
	60.4	

5G In the Keyes process for making absolute alcohol from alcohol containing 5% by weight of water, a third component benzene is added to the alcohol feed. Benzene lowers the volatility of alcohol and takes the entire water impurity overhead as a constant boiling mixture of composition 18.5% alcohol, 7.5% water and 74% benzene. The residue from the still is absolute alcohol. Compute the amount of benzene and impure alcohol required for the production of 2 500 kg absolute alcohol.

Solution

Basis: 2 500 kg of absolute alcohol

The process is represented diagrammatically in Figure 5.4. Let x_A be the weight fraction of alcohol in feed and x_W be the weight fraction of water in feed. Feed F fed to the distillation column is a mixture of benzene and impure alcohol, containing 5% water. The distillate D contains 18.5% alcohol and 7.5% water. The residue R is absolute alcohol.

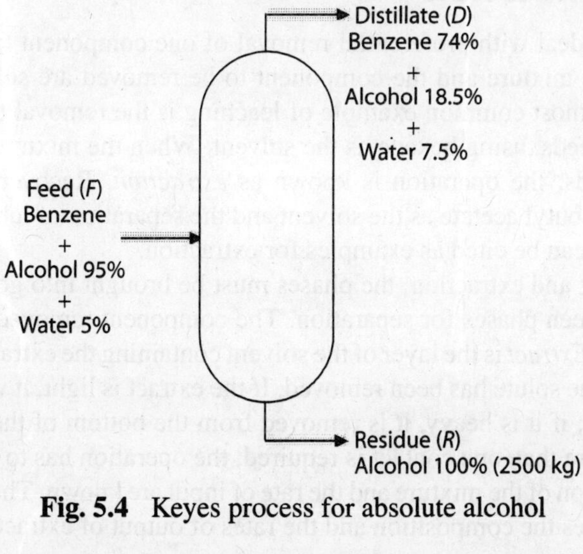

Fig. 5.4 Keyes process for absolute alcohol

Alcohol balance

Alcohol in feed = absolute alcohol + alcohol in the distillate

$$F x_A = 2\,500 + 0.185\,D \quad (1)$$

Water balance

Water in feed = water in the distillate

$$F x_W = 0.075\,D \quad (2)$$

$$x_A = 19\,x_W \quad (3)$$

From (1), (2) and (3)

$$19 \times 0.075\,D = 2\,500 + 0.185\,D$$

$$D = 2\,016.13.$$

Distillate plus residue is equal to feed.

Feed $= 2\,500 + 2\,016.13 = 4\,516.13$ kg

74% of the distillate is benzene.

Benzene used $= 2\,016.13 \times \dfrac{74}{100} = 1\,491.94$ kg

Feed is made by mixing benzene and impure alcohol.

Impure alcohol used = feed − benzene $= 4\,510.13 - 1\,491.94 = 3\,024.19$ kg

LEACHING AND EXTRACTION

Leaching and extraction deal with preferential removal of one component from its mixture by a liquid solvent. When the mixture and the component to be removed are solids, the operation is known as *leaching*. The most common example of leaching is the removal of edible oil from oil cake obtained from oil seeds, using hexane as the solvent. When the mixture and the component to be removed are liquids, the operation is known as *extraction*. Recovery of penicillin from fermentation broth using butyl acetate as the solvent and the separation of lubricating oil fractions using furfural as solvent can be cited as examples for extraction.

In the case of leaching and extraction, the phases must be brought into good contact to permit transfer of material between phases for separation. The component removed from the mixture is usually known as solute. *Extract* is the layer of the solvent containing the extracted solute. *Raffinate* is the layer from which the solute has been removed. If the extract is light, it will be removed from the top of the equipment; if it is heavy, it is removed from the bottom of the equipment. For the desired separation, if more than one contact is required, the operation has to be repeated.

Usually, the composition of the mixture and the rate of input are known. The extent of separation of the solute aimed at fixes the composition and the rates of output of extract and raffinate.

ILLUSTRATIONS

5H 2 000 kg h^{-1} of a slurry containing 20% suspended solids and the rest water is fed to a thickener. The overflow from the thickener is free from suspended solids, while the underflow contains 40% solids. The underflow is subsequently fed to a filter. The filter is free from suspended solids, while the filter cake contains 80% solids and 20% water. Compute the rates of underflow, overflow, filtrate and filter cake.

Solution

Basis: 1 hour operation

The process is represented diagrammatically in Figure 5.5.

Weight of feed $= 2\,000$ kg

Solids in the feed $= 2\,000 \times 0.2 = 400$ kg

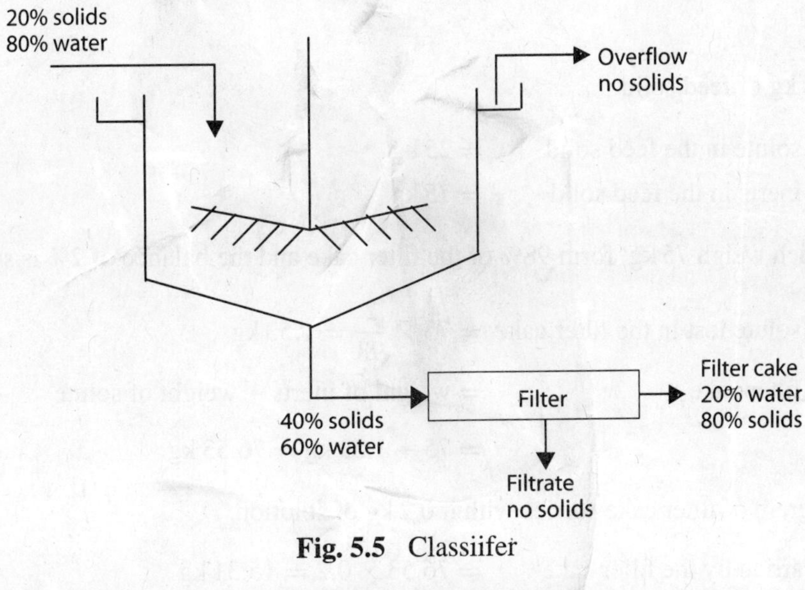

Fig. 5.5 Classiifer

The thickener has one input (the feed), and two outputs (overflow and underflow). The overflow has no solids. So all the solids entering the thickener, should leave as underflow. The underflow contains 40% solids, which should be 400 kg.

$$\text{Weight of underflow} = 400 \times \frac{100}{40} = 1000\,\text{kg h}^{-1}$$

Hence the weight of overflow = weight of feed − weight of underflow

$$= 2\,000 - 1000 = 1000\,\text{kg h}^{-1}$$

The underflow is fed to the filter. The filtrate has no solids, while the filter cake contains 80% solids, which is 400 kg.

Weight of filler cake $= 400 \times \dfrac{100}{80} = 500\,\text{kg h}^{-1}$

Weight of filtrate = weight of underflow from thickener − weight of filter cake

$= 1000 - 500 = 500\,\text{kg h}^{-1}$

5I In a leaching unit, feed solid containing 25% solute and 75% inerts is treated with fresh solvent. After allowing sufficient time, the contents of the unit are filtered. The filtrate contains 10% solute and 90% solvent. The filter cake contains 2% solid solute and it carries with it 0.2 kg of solution per kilogram of solids. Compute (i) the percentage recovery of solute, and (ii) the weight of solvent used for this operation.

Solution

Basis: 100 kg of feed solid

Weight of solute in the feed solid = 25 kg

Weight of inerts in the feed solid = 75 kg

Inerts, which weigh 75 kg, form 98% of the filter cake and the balance of 2% is solute.

Weight of solute lost in the filter cake $= 75 \times \dfrac{2}{98} = 1.53\,\text{kg}$

Weight of filter cake = weight of inerts + weight of solute

$= 75 + 1.53\,\text{kg} = 76.53\,\text{kg}$

Each kilogram of filter cake carries with it 0.2 kg of solution.

Solution carried by the filler cake $= 76.53 \times 0.2 = 15.31\,\text{kg}$

This solution has the same composition as the filtrate. The filtrate contains 10% solute and 90% solvent.

Weight of solute in the solution carried by the filter cake $= 15.31 \times \dfrac{10}{100} = 1.53\,\text{kg}$

Weight of solvent in the solution carried by the filler cake = weight of solution − weight of solute

$= 15.31 - 1.53 = 13.78\,\text{kg}$

$$\text{Weight of total solute lost in cake} = \frac{\text{weight of solute}}{\text{in the cake}} + \frac{\text{weight of solute}}{\text{in the solution}}$$

$$= 1.53 + 1.53 = 3.06\,\text{kg}$$

$$\text{Weight of solute recovered} = \frac{\text{weight of solute}}{\text{in feed}} - \frac{\text{weight of solute}}{\text{lost in filter cake}}$$

$$= 25 - 3.06 = 21.94\,\text{kg}$$

$$\text{Percentage recovery of solute} = \frac{21.94}{25} \times 100 = 87.76\%$$

The filtrate contains 10% solute and 90% solvent.

$$\text{Weight of solvent in the filtrate} = 21.94 \times \frac{90}{10} = 197.46\,\text{kg}$$

$$\text{Weight of solvent lost in filter cake} = 13.78\,\text{kg}$$

$$\text{Weight of solvent used in the operation} = \frac{\text{weight of solvent}}{\text{in the filtrate}} + \frac{\text{weight of solvent}}{\text{lost in the filter cake}}$$

$$= 197.46 + 13.78 = 211.24\,\text{kg per100 kg of feed solid}$$

5J Methylethylketone (MEK) and water are partially miscible and hence form a mutually soluble two phase system. At 298 K, the MEK-rich layer contains 12% water and the water-rich layer contains 25% MEK. If 100 kg each of water and MEK are mixed at 298 K and allowed to reach equilibrium, estimate the weight of the two layers.

Solution

Basis: 100 kg of water mixed with 100 kg of MEK at 298 K

On mixing, two layers, one rich in MEK and the other rich in water are formed.

Total weight of the two layers = 200 kg

Let the weight of the MEK-rich layer be x kg. Then the weight of the water-rich layer will be $(200 - x)$ kg. We have one unknown, x, and to find its value one equation is required. The required equation is framed from a water balance.

Water balance

Weight of water in the MEK-rich layer + weight of water in the water-rich layer = weight of total water

The MEK-rich layer contains 12% water and the water-rich layer contains 25% MEK.

$$0.12\,x + 0.75(200 - x) = 100$$

$$\therefore \quad x = 79.37\,\text{kg}$$

Weight of MEK-rich layer = 79.37 kg
Weight of water-rich layer = 200 – 79.37 = 120.63 kg

5K A mixture of phenol and water under certain conditions of temperature and composition forms two separate liquid phases, one rich in phenol and the other rich in water. At 303 K, the composition of phenol in the upper and lower layers is 70% and 9% respectively. If 40 kg of phenol and 60 kg of water are mixed, and the layers are allowed to separate at 303 K, what will each layer weigh?

Solution

Basis: Mixture of 40 kg phenol and 60 kg water

The process is represented diagrammatically in Figure 5.6

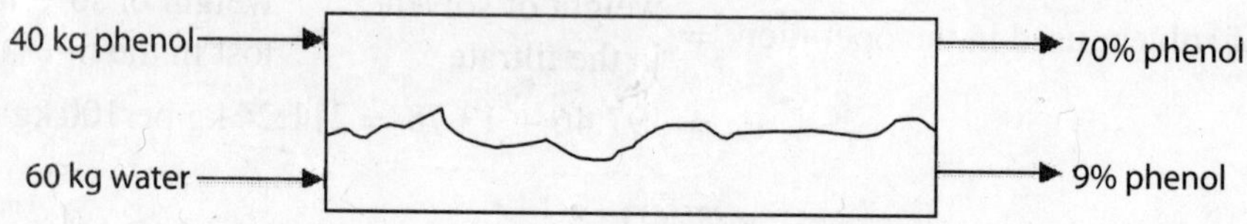

Fig. 5.6 Mixing of phenol and water

Total weight of the two layers is 100 kg. Let the weight of the upper layer be x kg. Then the weight of the lower layer is $(100 - x)$ kg. We have one unknown, x, and to find its value, one equation is required. The required equation is framed from a phenol balance.

Phenol forms 70% of the upper layer and 9% of the lower layer.

Phenol balance

Phenol in upper layer
+ phenol in lower layer = total phenol

$0.7x + 0.09(100 - x) = 40$

$\therefore \quad x = 50.82$ kg

Weight of upper layer = 50.82 kg

Weight of lower layer = 100 – 50.82 = 49.18 kg

EVAPORATION

Evaporation is a unit operation in which a solution containing a volatile solvent and a non-volatile solute is concentrated by evaporating a portion of the solvent. The solvent is mostly water and thick liquor is the product. Evaporators are heated by steam condensing on the metallic tubes. Nearly always, the solution to be evaporated flows inside the tubes.

Evaporation differs from drying in that the residue is a liquid—sometimes a highly viscous one—rather than a solid; it differs from distillation in that the vapour is usually a single component; it differs from crystallisation in that the emphasis is placed on concentrating a solution rather than forming and building crystals.

Normally, in evaporation, the vapour from the boiling liquid is condensed and discarded. This method is called single effect evaporation and it utilises steam ineffectively. To evaporate 1 kg of water from a solution, 1 to 1.3 kg of steam is required. If the vapour from one evaporator is fed to the steam chest of a second evaporator, and the vapour from the second is then sent to a condenser, the operation becomes double effect. The heat in the original steam is reused in the second effect, and the evaporation achieved by unit mass of steam fed to the first effect is approximately doubled. Additional effects can be added in the same manner.

In the problems on evaporation, the composition of the feed and the feed rate are generally given. The composition of the thick liquor determines the rate of production of thick liquor and the rate of steam generation.

ILLUSTRATIONS

5L A spent lye sample obtained from a soap making unit contains 14% glycerol and 12% sodium chloride by weight. It is concentrated at the rate of 6 000 kg h^{-1} in a triple effect evaporator until the final solution contains 85% glycerol and 5% sodium chloride by weight. Assuming that 6% of glycerol is lost by entrainment, calculate (i) the amount of evaporation taken place in the system, and (ii) the amount of salt crystallised out in the salt box of the evaporator.

Solution

Basis: 6 000 kg h^{-1} lye fed to the evaporator

The process is represented diagrammatically in Figure 5.7.

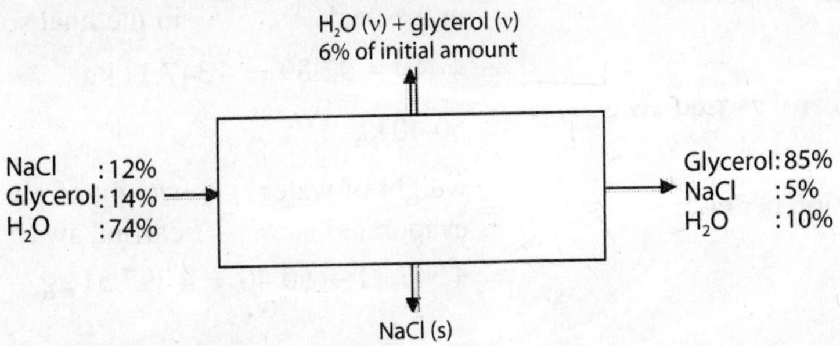

Fig. 5.7 Single effect evaporator

The feed contains 14% glycerol and 12% sodium chloride by weight

$$\text{Weight of glycerol in the feed} = \frac{14}{100} \times 6\,000 = 840\,\text{kg}$$

Weight of sodium chloride in the feed $= \dfrac{12}{100} \times 6\,000 = 720\,\text{kg}$

Weight of water in the feed $= 6\,000 - 840 - 720 = 4\,440\,\text{kg}$

6% of glycerol is carried away by the water vapour evaporated.

Weight of glycerol lost by entrainment $= 840 \times \dfrac{6}{100} = 50.4\,\text{kg}$

The remaining glycerol will be present in the final solution.

Weight of glycerol in the final solution $= 798.6\,\text{kg}$

Final solution contains 85% glycerol, 5% NaCl and 10% water by weight.

Weight of glycerol in the final solution $= 798.6\,\text{kg}$

Weight of sodium chloride in the final solution $= 789.6 \times \dfrac{5}{85} = 46.45\,\text{kg}$

Weight of water in the final solution $= 789.6 \times \dfrac{10}{85} = 92.89\,\text{kg}$

For sodium chloride, the final solution is one output, while the other output is salt crystallised in the salt box.

Weight of sodium chloride crystallised out = weight of sodium chloride in the feed − weight of sodium chloride in the final solution

$= 720 - 46.45 = 673.55\,\text{kg}$

For water, final solution is one output, while the other output is water evaporated.

Weight of water evaporated = weight of water in the feed − weight of water in the final solution

$= 4\,440 - 92.89 = 4\,347.11\,\text{kg}$

Weight of glycerol carried away by water vapour $= 50.40\,\text{kg}$

Total evaporation per hour = weight of water evaporated + weight of glycerol carried away

$= 4\,347.11 + 50.40 = 4\,397.51\,\text{kg}$

5M $5\,000\,\text{kg h}^{-1}$ of an aqueous sodium hydroxide solution containing 10wt% of sodium hydroxide is concentrated to 50wt% sodium hydroxide in a double effect evaporator. The feed solution is partially concentrated in the first evaporator and then fed to the second evaporator. Find the amount of 50wt% sodium hydroxide solution produced per hour, and the amount of water evaporated per day. Assuming that the water evaporating capacity of both the evaporators to be the same, find the concentration of sodium hydroxide solution leaving the first evaporator.

Solution

Basis: 1 hour operation

The process is represented diagrammatically in Figure 5.8.

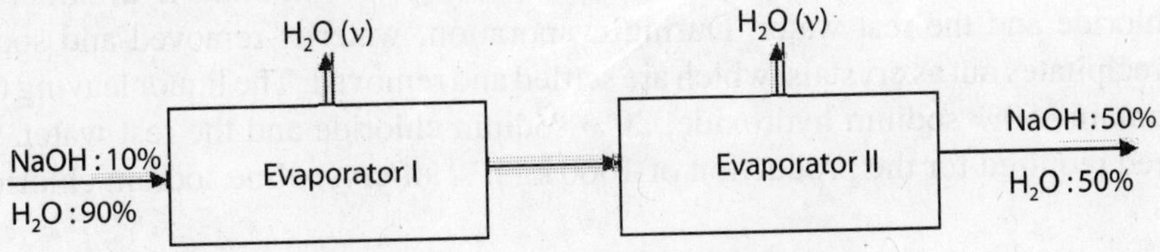

Fig. 5.8 Double effect evaporator

Weight of feed $= 5\,000$ kg

Feed contains 10wt% sodium hydroxide and the balance is water.

Weight of sodium hydroxide in the feed $= 5\,000 \times \dfrac{10}{100} = 500$ kg

Weight of water in the feed $= 5\,000 - 500 = 4\,500$ kg

For sodium hydroxide there is only one output. So all the sodium hydroxide should be present in that output.

$\therefore$ Sodium hydroxide in the final solution $= 500$ kg

This forms 50% of the final solution.

Weight of final solution produced $= 500 \times \dfrac{100}{50} = 1000 \text{ kg h}^{-1}$

Weight of water in the final solution $=$ total weight $-$ weight of sodium hydroxide

$= 1000 - 500 = 500$ kg

Weight of water evaporated $=$ water in feed $-$ water in final solution.

$= 4\,500 - 500 = 4\,000 \text{ kg h}^{-1}$

If the water evaporated in both the evaporators is equal, it will be 2 000 kg in each. So the composition of the solution leaving the first evaporator will be:

Weight of sodium hydroxide $= 500$ kg

Weight of water $=$ water entering the first effect $-$ water evaporated in the first effect

$= 4\,500 - 2\,000 = 2\,500$ kg

Weight of the solution output from first evaporator $= \text{weight of water} + \text{solute}$

$= 2\,500 + 500 = 3\,000\,\text{kg}$

Weight per cent of sodium hydroxide in the solution output from the first evaporator $= \dfrac{500}{3\,000} \times 100 = 16.67$

5N An evaporator is fed with a solution containing 10% sodium hydroxide, 10% sodium chloride and the rest water. During evaporation, water is removed and sodium chloride precipitates out as crystals, which are settled and removed. The liquor leaving the evaporator contains 50% sodium hydroxide, 20% sodium chloride and the rest water. Calculate the feed required for the production of 1000 kg h^{-1} of crystalline sodium chloride.

Solution

Basis: 100 kg of feed

The process is represented diagrammatically in Figure 5.9.

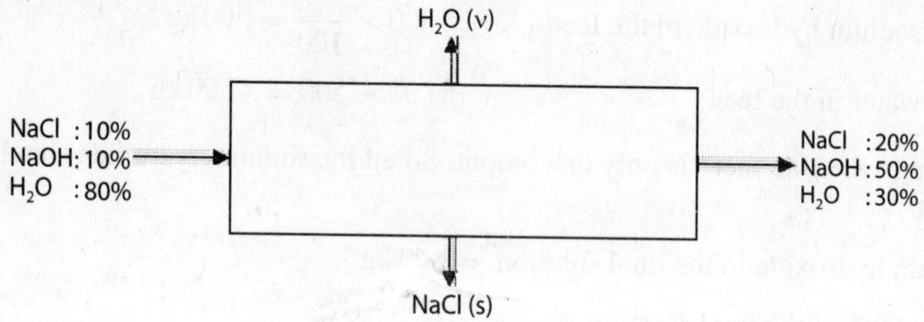

Fig. 5.9 Single effect evaporator

Sodium hydroxide in the feed $= 100 \times \dfrac{10}{100} = 10\,\text{kg}$

Sodium chloride in the feed $= 100 \times \dfrac{10}{100} = 10\,\text{kg}$

For sodium hydroxide there is only one output. So all sodium hydroxide should leave by that.

Sodium hydroxide in the final solution $= 10\,\text{kg}$

This forms 50% of the final solution.

Weight of final solution produced $= 10 \times \dfrac{100}{50} = 20\,\text{kg}$

Final solution contains 20% sodium chloride.

Sodium chloide in final solution $= 20 \times \dfrac{20}{100} = 4.0\,\text{kg}$

For sodium chloride, the final solution is one output, while the other output is salt crystallised.

$$\text{Sodium chloride crystallised out} = \begin{matrix}\text{sodium chloride}\\ \text{in the feed}\end{matrix} - \begin{matrix}\text{sodium chloride}\\ \text{in the final solution}\end{matrix}$$

$$= 10 - 4.0 = 6.0\,\text{kg}$$

For 6.0 kg sodium chloride production, feed solution required $= 100\,\text{kg}$

For 1000 kg sodium chloride production, feed solution required $= \dfrac{100}{6.0} \times 1000 = 16\,666.67\,\text{kg h}^{-1}$

DISSOLUTION AND CRYSTALLISATION

Dissolving a solute in a solvent to produce a homogenous solution of the solute in the solvent is known as *dissolution*. When the solvent cannot dissolve any more solute, the additional solute added will remain undissolved, and the resultant solution is known as a *saturated solution*. If a saturated solution is heated, more solute will go into solution; if it is cooled, part of the dissolved solute will crystallise out.

Crystallisation is a unit operation in which the dissolved solids of a solution are separated by the dependence of solubility on temperature or concentration. If crystallisation is based on the difference in solubility with temeperature, the feed solution is cooled. On the other hand, if the crystallisation is based on the difference in solubility with concentration, a portion of the solvent in the feed solution is evaporated. If evaporation followed by cooling is done, an enhanced yield can be obtained.

The feed solution may or may not be saturated at the feed temperature; but it will be *supersaturated* at the crystallisation temperature.

The solution left behind after crystallisation is complete, is known as the *mother liquor*. The mother liquor is saturated at the crystallisation temperature. The excess solute over and above that

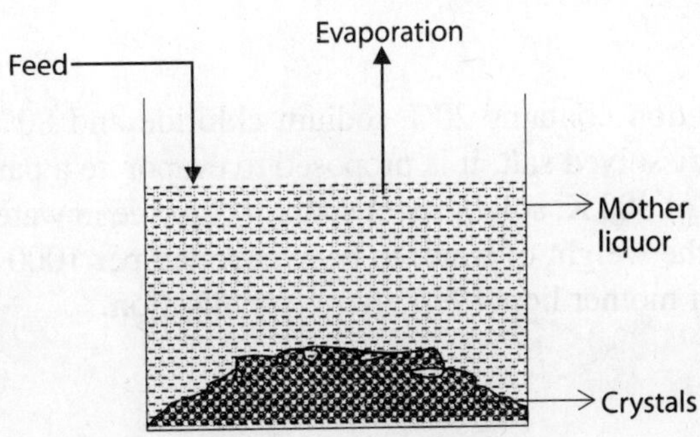

Fig. 5.10 Crystalliser

required to form a saturated solution at the crystallisation temperature is thrown out of the feed solution and is collected as *crystals*.

Yield of crystals = weight of solute in the feed

−weight of solute retained in the mother liquor

Let the solubility of A in B be 50 kg per 100 kg B at 293 K and 80 kg per 100 kg B at 323 K. If the saturated solution at 323 K is cooled to 293 K, the amount of A crystallized will be 30 kg.

Problems on crystallisation can be easily solved based on material balance equations: overall balance and solute balance.

1. Input is the feed solution (F). If crystallisation is accompanied by evaporation and cooling, the outputs are mother liquor left (M), crystals formed (C) and evaporation loss (E).

   Overall balance

   $$F = M + C + E$$

   If there is no evaporation

   $$F = M + C \quad (1)$$

2. Let x_F be the weight fraction of solute in the feed, x_M be the weight fraction of solute in the mother liquor, and x_C be the weight fraction of solute in the crystals.

   Solute balance

   $$Fx_F = Mx_M + Cx_C \quad (2)$$

   It may be noted that x_C is the ratio of the molecular weight of anhydrous solute to that of the hydrated solute.

   ∴ For hydrated solutes (solutes with water of crystallisation), $x_C < 1$ and for anhydrous solutes (solutes with no water of crystallisation), $x_C = 1$.

ILLUSTRATIONS

50 A common salt solution contains 20% sodium chloride and 80% water by weight. To recover 90% of the dissolved salt, it is proposed to evaporate a part of the water and then crystallise at 303 K. At 303 K, solubility of sodium chloride in water is 40 kg per 100 kg of water. Calculate (i) the weight of water to be evaporated per 1000 kg of original solution, and (ii) the weight of mother liquor left after crystallisation.

Solution

Basis: 1000 kg of solution

Weight of sodium chloride in the solution $= 1000 \times \dfrac{20}{100} = 200\,\text{kg}$

Weight of water in the solution $= 800\,\text{kg}$

90% of sodium chloride is to be recovered as crystals.

Weight of sodium chloride to be crystallised $= 200 \times \dfrac{90}{100} = 180\,\text{kg}$

Sodium chloride in mother liquor $=$ sodium chloride in feed $-$ sodium chloride crystallised out

$= 200 - 180 = 20\,\text{kg}$

The mother liquor is saturated at the crystallisation temperature at which the solubility of sodium chloride in water is 40 kg per 100 kg of water.

To retain 20 kg of sodium chloride in solution the amount of water required $= 20 \times \dfrac{100}{40} = 50\,\text{kg}$

Weight of mother liquor $=$ weight of water $+$ weight of sodium chloride

$= 50 + 20 = 70\,\text{kg}$

Water to be evaporated $=$ water in feed $-$ water in mother liquor

$= 800 - 50 = 750\,\text{kg}$

5P After a crystallisation process, a solution of calcium chloride in water contains 50 parts of calcium chloride per 100 parts of water. Calculate the weight of this solution necessary to dissolve 200 kg of $CaCl_2.6H_2O$ crystals at a temperature of 298 K. Solubility of calcium chloride in water at 298 K is 7.38 kmol of calcium chloride per 1000 kg of water.

Solution

Basis: 200 kg of $CaCl_2.6H_2O$ solute

Let the solution required to dissolve the above solute be x kg. The solution contains 50 parts of calcium chloride per 100 parts of water.

Calcium chloride, already present in the solution $= \dfrac{50x}{(100+50)} = 0.333\,3x$

Solvent water in the solution $= 0.666\,7x$

Solute to be dissolved $= 200\,\text{kg}$

Anhydrous calcium chloride in the solute $= 200 \times \dfrac{111}{219} = 101.37\,\text{kg}$

(219 is the molecular weight of hydrated salt and 111 is the weight of calcium chloride in it)

Water of crystallisation in the solute $= 200 - 101.37 = 98.63$ kg

Total amount of water $=$ water of crystallisation in the solute $+$ water in the solution

$= 98.63\text{ kg} + 0.6667x.$

Total calcium chloride $=$ calcium chloride in the solute $+$ calcium chloride in the solution

$= 101.37\text{ kg} + 0.3333x$

Solubility at 298 K $= 7.38$ kmol calcium chloride per 1000 kg water

In the final solution, ratio of calcium chloride to water $= 7.38 \times \dfrac{111}{1000} = \dfrac{819.2}{1000}$

$$\frac{\text{Solute}}{\text{Water}} = \frac{819.2}{1000} = \frac{0.3333x + 101.37}{0.6667x + 98.63}$$

$\therefore$ Weight of solution required (x) $= 96.65$ kg

5Q A solution of potassium dichromate in water contains 15% potassium dichromate by weight. 1000 kg of this solution is evaporated to remove some amount of water. The remaining solution is cooled to 298 K. If the yield of potassium dichromate crystals is 75%, calculate the amount of water evaporated. Solubility of potassium dichromate in water is 0.39 kmol per 1000 kg of water at 298 K.

Solution

Basis: 1000 kg of 15% potassium dichromate solution

Weight of potassium dichromate in the solution $= 150$ kg

Weight of water in the solution $= 1000 - 150 = 850$ kg

75% of the solute has to be crystallised out.

Potassium dichromate to be crystallised $= 150 \times \dfrac{75}{100} = 112.5$ kg

The balance will remain in solution. The solution remaining after crystallisation is the mother liquor.

Potassium dichromate in mother liquor $= 150 - 112.5 = 37.5$ kg

The mother liquor is saturated at the crystallisation temperature at which the solubility of potassium dichromate in water is 0.39 kmol per 1000 kg of water.

Solubility of potassium dichromate $= 0.39 \times 294$ kg per 1000 kg of water

$= 114.66$ kg per 1000 kg water

Water required to retain 37.5 kg solute in mother liquor $= 1000 \times \dfrac{37.5}{114.66}$

$= 327.05$ kg.

The remaining water has to be removed by evaporation.

Weight of water to be evaporated = water in the feed – water in mother liquor

$= 850.00 - 327.05 = 522.95$ kg

5R 500 kg of 20wt% solution of sodium nitrate is concentrated to saturation by evaporation. The solution is then cooled to 293 K. The crystals of sodium nitrate are removed by filtration. The wet crystals carry 5% by weight of mother liquor and are dried. The solubility of sodium nitrate is 1.76 kg per kilogram of water at 373 K, and 0.88 kg per kilogram of water at 293 K. Calculate (i) the amount of water evaporated to achieve saturation at 373 K, and (ii) the total weight of sodium nitrate obtained.

Solution

Basis: 500 kg of feed solution

Weight of sodium nitrate in the feed solution $= \dfrac{20}{100} \times 500 = 100$ kg

Weight of water in the feed solution $= 500 - 100 = 400$ kg

The feed solution is saturated by evaporation at 373 K. The solubility of sodium nitrate is 1.76 kg per kilogram of water at 373 K. At 373 K, 1 kg of water is required for 1.76 kg of sodium nitrate, to make a saturated solution.

Weight of water required for saturated solution at 373 K $= 100 \times \dfrac{1}{1.76} = 56.82$ kg

The remaining water has to be evaporated.

Weight of water to be evaporated = water in feed – water in saturated solution

$= 400 - 56.82 = 343.18$ kg

The saturated solution at 373 K is cooled to 293 K. The resultant solution is saturated at 293 K. As the temperature has come down, the solubility also will come down and the excess solute will
10 crystallise out. The solution remaining after the crystallisation is complete is the mother liquor. In

the present case there is no evaporation. So water in the original solution and the mother liquor is the same.

Water in the mother liquor $= 56.82\,\text{kg}$

The mother liquor is saturated at the crystallisation temperature at which the solubility of sodium nitrate is 0.88 kg per kg of water.

Sodium nitrate in the mother liquor $= 56.82 \times 0.88 = 50\,\text{kg}$

Sodium nitrate crystallised out $= \begin{matrix}\text{sodium nitrate in} \\ \text{the original solution}\end{matrix} - \begin{matrix}\text{sodium nitrate in the} \\ \text{mother liquor}\end{matrix}$

$= 100 - 50 = 50\,\text{kg}$

Weight of mother liquor $= \text{weight of solvent} + \text{weight of solute}$

$= 56.82 + 50.00 = 106.82\,\text{kg}$

Wet crystals carry 5% of the mother liquor.

Mother liquor wetting the crystals $= 106.82 \times \dfrac{5}{100} = 5.34\,\text{kg}$

Weight of sodium nitrate in 106.82 kg of mother liquor $= 50\,\text{kg}$

Weight of sodium nitrate in 5.34 kg of mother liquor $= 5.34 \times \dfrac{50}{106.82} = 2.5\,\text{kg}$

Total weight of sodium nitrate crystals $= \begin{matrix}\text{sodium nitrate} \\ \text{crystalised out}\end{matrix} + \begin{matrix}\text{sodium nitrate from mother liquor} \\ \text{wetting the crystals liquor}\end{matrix}$

$= 50 + 2.5 = 52.5\,\text{kg}$

5S 2 000 kg of 30% solution of sodium carbonate are cooled slowly to 293 K. During cooling, 5% of the solution is evaporated. The carbonate precipitating out forms $Na_2CO_3.10H_2O$ crystals. The solubility of anhydrous sodium carbonate at 293 K is 21.5 kg per 100 kg of water. What is the weight of decahydrate crystals, water loss on evaporation and mother liquor left?

Solution

Basis: 2 000 kg of feed (F)

Weight fraction of sodium carbonate solute in the feed (x_F) $= \dfrac{30}{100} = 0.3$

Weight of sodium carbonate solute in the feed $= 2\,000 \times 0.3 = 600\,\text{kg}$

Weight of water in the feed $= 2\,000 - 600 = 1\,400$ kg

During cooling, 5% of the solution is evaporated.

Weight of water evaporated (E) $= 2\,000 \times \dfrac{5}{100} = 100$ kg

Mother liquor is saturated at the crystallisation temperature, 293 K. Solubility of anhydrous sodium carbonate at 293 K is 21.5 kg per 100 kg of water.

Weight fraction of sodium carbonate in the mother liquor (x_M) $= \dfrac{21.5}{(21.5 + 100)} = 0.177$

The composition of crystals (C) is $Na_2CO_3.10\,H_2O$. Molecular weight of anhydrous Na_2CO_3 is 106 and that of hydrated crystals is 286.

Weight fraction of Na_2CO_3 in the hydrated crystals (x_C) $= \dfrac{106}{286} = 0.371$

The problem can be solved using the overall balance and solute balance.

Overall balance

$$F = M + C + E$$

$$2\,000 = M + C + 100 \qquad (1)$$

Solute balance

$$F x_F = M x_M + C x_C$$

$$600 = 0.177M + 0.371\,C \qquad (2)$$

From (1) and (2)

Weight of mother liquor $= 540.72$ kg

Weight of crystals $= 1359.28$ kg

5T In a particular process, 2 000 kg of crystalline $Na_2SO_4.10\,H_2O$ are obtained and the mother liquor was found to contain 20% Na_2SO_4 (anhydrous) by weight. If the feed solution contained 30% Na_2SO_4 (anhydrous) by weight and 20% of original water is lost by evaporation, calculate the weight of feed solution, weight of mother liquor left and weight of water evaporated.

Solution

Basis: As the amount of feed is not known, we have to assume the feed to be 100 kg, and find the crystals obtainable. From that, we have to back calculate the other unknowns.

Weight fraction of solute in feed (x_F) $= 0.3$

Weight of Na_2SO_4 in 100 kg feed $= 30$ kg

Weight of water in 100 kg feed $= 70$ kg

20% of original water is lost by evaporation.

Water evaporated (E) $= 70 \times \dfrac{20}{100} = 14$ kg

Weight fraction of solute in mother liquor (x_M) $= 0.2$

Molecular weight of Na_2SO_4 is 142 and that of $Na_2SO_4.10\,H_2O$ is 322.

Weight fraction of anhydrous solute in the crystals (x_C) $= \dfrac{142}{322} = 0.441$

The problem can be solved using the overall balance and solute balance.

Overall balance

$$F = M + C + E$$

$$100 = M + C + 14$$

$$86 = M + C \qquad (1)$$

Solute balance

$$Fx_F = Mx_M + Cx_C$$

$$30 = 0.2\,\text{M} + 0.441\,\text{C} \qquad (2)$$

From (1) and (2)

$$C = 53.11 \text{ kg}$$

If the weight of crystals is 53.11 kg, weight of feed is 100 kg.

If the crystals weight 2 000 kg, weight of feed $= \dfrac{100}{53.11} \times 2\,000 = 3\,765.77$ kg

If the weight of feed is 100 kg, evaporation loss $= 14$ kg

If the weight of feed is 3 756.77 kg, evaporation loss $= \dfrac{14}{100} \times 3\,765.77 = 527.2$ kg

Weight of mother liquor left = weight of feed − weight of crystals − weight of evaporation

$$= 3\,756.77 - 2\,000 - 527.2$$

$$= 1\,238.57 \text{ kg}$$

5U Solubility of $FeSO_4.7\,H_2O$ is 47.5 kg and 33.8 kg in 100 kg of water at 318 K and 305 K respectively. (i) Estimate the theoretical weight of saturated solution at 318 K to be charged to get 1000 kg of crystals by cooling to 305 K. (ii) If 5% of total water entering the system is lost by evaporation, what is the yield of crystals in this case? (ii) At the conditions of (ii), how much of feed solution is required for a harvest of 1000 kg of crystals?

Solution

Basis: 147.5 kg of saturated solution at 318 K

(i) Solubility at 318 K = 47.5 kg

Solubility at 305 K = 33.8 kg

If a saturated solution at 318 K is cooled to 305 K, yield of crystals will be equal to the difference in solubility.

Yield of crystals = 13.7 kg

For 13.7 kg crystals, the weight of feed required = 147.5

For 1000 kg crystals, the weight of feed required $= 147.5 \times \dfrac{100}{13.7} = 10\,766.4$ kg

(ii) 147.5 kg feed contains 100 kg water.

Weight of water in 10 766.4 kg feed $= 10\,766.4 \times \dfrac{100}{147.5} = 7\,299.3$ kg

During cooling 5% of water is lost by evaporation.

Water lost by evaporation $= 7\,299.3 \times \dfrac{5}{100} = 364.96$ kg

Solute dissolved in this water will be available as additional crystals.
Solubility at 305 K is 33.8 kg in 100 kg of water.

Additional yield $= 364.96 \times \dfrac{33.8}{100} = 123.4$ kg

Total yield = original yield + additional yield = 1 123.4 kg

(iii) Under conditions of (ii),

Weight of yield required for a feed of 1 123.4 kg $= 10\,766.4$ kg

Weight of feed, if the yield required is only 1000 kg $= 10\,766.4 \times \dfrac{1000}{1\,123.4} = 9\,584.6$ kg

5V A hot solution of barium nitrate from an evaporator contains 30.6 kg barium nitrate per 100 kg of water and goes to a crystalliser where the solution is cooled and barium nitrate crystallises. On cooling 10% of the original water evaporates. For a feed of 100 kg, calculate (i) the yield of crystals if the solution is cooled to 290 K, where the solubility is 8.5 kg of barium nitrate per 100 kg water, and (ii) the yield of crystals if the solution is cooled to 280 K, where the solubility is 6.9 kg of barium nitrate per 100 kg of water.

Solution

Basis: 100 kg of feed

(i) Weight fraction of solute in feed (x_F) $= \dfrac{30.6}{(30.6 + 100)} = 0.234$

Weight fraction of water in the feed $= 1 - 0.234 = 0.766$

Weight fraction of solute in mother liquor (x_M) $= \dfrac{8.5}{(8.5 + 100)} = 0.078$

The crystals are not hydrated.

Weight fraction of anhydrous solute in the crystals obtained (x_C) $= 1.0$

Weight of water in the feed $= 100 \times 0.766 = 76.6$ kg

10% of water is lost during cooling.

Weight of water evaporated (E) $= 76.6 \times \dfrac{10}{100} = 7.66$ kg

The problem can be solved using the overall balance and solute balance.

Overall balance

$$F = M + C + E$$

$$100 = M + C + 7.66 \quad (1)$$

Solute balance

$$Fx_F = Mx_M + Cx_C$$

$$23.4 = 0.078M + C \quad (2)$$

From (1) and (2)

$$\text{Yield } (C) = 17.48 \text{ kg}$$

(ii) If the solution is cooled to 280 K,

$$\text{Weight fraction of solute in mother liquor } (x_M) = \frac{6.9}{(6.9 + 100)} = 0.065$$

New solute balance is

$$23.4 = 0.065M + C \tag{3}$$

From (1) and (3)

$$\text{New yield } (C) = 18.53 \text{ kg}$$

5W A tank holds 5 000 kg of a saturated solution of sodium bicarbonate at 333 K. If 200 kg of sodium bicarbonate has to be crystallised from this solution, to what temperature must the solution be cooled.

Temperature (K)	Solubility of $NaHCO_3$ per 100 kg of water (kg)
333	16.4
323	14.45
313	12.7
303	11.1
293	9.6

Solution

Basis: 5 000 kg of a saturated solution of sodium bicarbonate at 333 K

Solubility at 333 K = 16.4 kg of sodium bicarbonate per 100 kg of water

$$\text{Weight fraction of solute} = \frac{16.4}{(100 + 16.4)} = 0.141$$

Sodium bicarbonate in the feed solution $= 5\,000 \times 0.141 = 705$ kg

Water in the feed solution $= 5\,000 - 705 = 4\,295$ kg

Sodium bicarbonate to be crystallised $= 200$ kg

We know that the solution from which crystallisation takes place is saturated at the crystallisation temperature.

Sodium bicarbonate in the remaining solution $= 705 - 200 = 505$ kg

Solubility of the solute at that stage $= 505 \times \dfrac{100}{4\,295}$

$= 11.76$ kg of bicarbonate per 100 kg of water

From the data given in the problem, this value is between the solubility values for the temperatures 303 K and 313 K.

Solubility at 313 K $= 12.7$ kg of bicarbonate per 100 kg H_2O

Solubility at 303 K $= 11.1$ kg of bicarbonate per 100 kg H_2O

Temperature to which the solution should be cooled is between 303 K and 313 K

The actual temperature is $= 303 + \left[10 \times \dfrac{(11.76 - 11.10)}{(12.70 - 11.10)}\right]$

$= 303 + 4.125 = 307.125$ K

5X Solubility of anhydrous manganese sulphate at 293 K is 62.9 g per 100 g of water. How much $MnSO_4.5H_2O$ must be dissolved in 100 kg of water to give a saturated solution?

Solution

Basis: 100 kg of water

Let x kg of $MnSO_4.5H_2O$ be required for making a saturated solution at 293 K.

Molecular weight of anhydrous manganese sulphate $= 151$

Molecular weight of $MnSO_4.5H_2O$ $= 241$

Weight of anhydrous manganese sulphate in x kg of $MnSO_4.5H_2O$ $= \dfrac{151x}{241} = 0.627x$

Weight of water of crystallisation in x kg of $MnSO_4.5H_2O$ $= 0.373x$

Total weight of water $= 100 + 0.373x$

Solute to solvent ratio should match with the solubility at 293 K.

$$\frac{\text{Weight of solute}}{\text{Weight of solvent}} = \frac{0.627x}{100 + 0.373x} = \frac{62.9}{100}$$

$$\therefore \quad x = 160.46\ \text{kg}$$

Solute required is 160.46 kg.

5Y The solubility of barium nitrate is 34 kg per 100 kg of water and 5 kg per 100 kg of water at 373 K and 273 K respectively. If you start with 100 kg of barium nitrate and make a saturated solution in water at 373 K, how much of water is required? If this solution is cooled to 273 K, how much of barium nitrate is precipitated out of the solution.

Solution

Basis: 100 kg of barium nitrate solute

A saturated solution at 373 K is prepared taking 100 kg of barium nitrate. The solubility of barium nitrate is 34 kg per 100 kg of water at 373 K, that is, 34 kg of solute requires 100 kg of water.

Water required for the dissolution of 100 kg of solute $= 100 \times \dfrac{100}{34} = 294.12\ \text{kg}$

As no evaporation takes place, the amount of water will remain the same and it is the water in the mother liquor.

Water in mother liquor $= 294.12\ \text{kg}$

The mother liquor is saturated at the crystallisation temperature, 273 K, when the solubility is 5 kg of barium nitrate per 100 kg water.

So 100 kg of water require 5 kg of solute.

Solute required by the water in mother liquor $= 5 \times \dfrac{294.12}{100} = 14.71\ \text{kg}$

The remaining solute will crystallise out.

Weight of crystals obtained $= 100 - 14.71 = 85.29\ \text{kg}$

5Z A solution of 53.8 gm of magnesium sulphate per 100 gm of water is cooled from 353 K to 323 K. During this process 6% of water gets evaporated. For 100 kg of original feed solution, calculate the amount of $MgSO_4.7H_2O$ crystals obtained and the quantity of mother liquor left. At 323 K, the mass fraction of magnesium sulphate should be 0.3.

Solution

Basis: 100 kg of feed solution (F)

Mass fraction of solute in feed (x_F) $= \dfrac{53.8}{(100 + 53.8)} = 0.35$

Weight of $MgSO_4$ in 100 kg of feed $= 100 \times 0.35 = 35$ kg

Weight of water in 100 kg of feed $= 100 - 35 = 65$ kg

During cooling 6% of water is evaporated.

Water evaporated (E) $= 65 \times \dfrac{6}{100} = 3.9$ kg

Molecular weight of $MgSO_4.7H_2O$ is 246.3 and that of anhydrous solute is 120.3.

Mass fraction of anhydrous solute in crystals (x_C) $= \dfrac{120.3}{246.3} = 0.488$

Mass fraction of solute in mother liquor (x_M) $= 0.3$

The problem can be solved using the overall balance and solute balance.

Overall balance

$$F = M + C + E$$

$$100 = M + C + 3.9$$

$$96.1 = M + C \qquad (1)$$

Solute balance

$$Fx_F = Mx_M + Cx_C$$

$$35 = 0.3M + 0.488C \qquad (2)$$

From (1) and (2), the yield of crystals (C) = 35.82 kg

SATURATION

It is a unit operation involving gas–liquid contact or gas–liquid mixtures. When a gas or gaseous mixture remains in contact with a liquid surface, the liquid will vapourise and the vapours will mix with the gas. This process will go on till the partial pressure of the vapour (p_v) in the gas–vapour mixture equals the equilibrium vapour pressure (P_v) of the liquid at that temperature. When the vapour composition reaches this equilibrium value, the gas is said to be saturated with the vapour. The vapour content of the saturated gas is determined entirely by vapour pressure of the liquid and may be predicted directly from vapour pressure–temperature data.

The vapour content of the gas may be expressed as:

1. volume per cent
2. weight per cent
3. weight of vapour per unit volume of vapour–gas mixture
4. weight of vapour per unit weight of vapour-free gas
5. moles of vapour per mole of vapour-free gas

ILLUSTRATION

5AA Acetone at a temperature of 295 K exerts a vapour pressure of 200 mm Hg. Calculate the composition of a saturated mixture of nitrogen and acetone vapour at 295 K and a pressure of 745 mm Hg expressed in the following terms: (i) percentage composition by volume, (ii) percentage composition by weight, (iii) kilograms of vapour per cubic metre of mixture, (iv) kilograms of vapour per kilogram of vapour-free gas, and (v) number of moles of vapour per mole of vapour-free gas.

Solution

Basis: 1 kmol of vapour–gas mixture

(i) Mole fraction is the ratio of partial pressure to total pressure.

Mole fraction of acetone in saturated acetone–nitrogen mixture $= \frac{200}{745} = 0.268$

Mole fraction of nitrogen in saturated acetone–nitrogen mixture $= 1 - 0268 = 0.732$

For gaseous mixtures, composition in mol% $\equiv$ composition in vol%

Mole per cent $=$ mole fraction $\times$ 100

Composition of the mixture in volume per cent

Percentage of acetone $= 26.8\%$
Percentage of nitrogen $= 73.2\%$

(ii) **Table 5.3** Composition of the mixture in weight per cent

Constituent	Mole fraction	Molecular weight	Weight (kg)	Weight per cent
Acetone	0.268	58	15.54	43.12
Nitrogen	0.732	28	20.50	56.88
			36.04	

Weight per cent of acetone $= 43.12\%$

Weight per cent of nitrogen $= 56.88\%$

(iii) Volume of 1 kmol at 745 mm and 293 K

$$= \frac{nRT}{P} = \frac{1 \times 0.08205 \times 293 \times 760}{745} = 24.53\ \text{m}^3$$

1 kmol of mixture contains 15.54 kg of acetone vapour.

24.53 m^3 of mixture contain 15.54 kg of acetone vapour.

Weight of acetone per cubic metre of mixture

$$= \frac{15.54}{24.53} = 0.634\ \text{kg}$$

(iv) 1 kmol of mixture contains 20.50 kg of nitrogen and 15.54 kg of acetone vapour.

Weight of vapour per kg of vapour-free gas

$$= \frac{15.54}{20.50} = 0.758\ \text{kg}$$

(v) 1 kmol of mixture contains 0.732 kmol of nitrogen and 0.268 kmol of acetone vapour.

Amount of vapour per mole of vapour-free gas

$$= \frac{0.268}{0.732} = 0.366\ \text{mol}$$

PARTIAL SATURATION

If a gas contains a vapour in such proportion that its partial pressure is less than the equilibrium vapour pressure of the liquid at that temperature, the mixture is partially saturated. The composition of a partially saturated gas–vapour mixture can be expressed by

- relative saturation
- percentage saturation

The *relative saturation* of a vapour–gas mixture is defined as the percentage ratio of the partial pressure of the vapour to the vapour pressure of the liquid at that temperature.

$$\text{Relative saturation } (Y_r) = (p_v/P_v) \times 100$$

$p_v =$ partial pressure of the vapour

$P_v =$ vapour pressure of the liquid

The *percentage saturation* of a vapour–gas mixture is defined as the percentage ratio of the existing number of moles of vapour per mole of vapour-free gas to the number of moles of vapour that would be present per mole of vapour-free gas if the mixture was saturated at the existing temperature and pressure.

$$\text{Percentage saturation } (Y_p) = (n_v/n_s) \times 100$$

n_v = number of moles of vapour per mole of vapour-free gas actually present

n_s = number of moles of vapour per mole of vapour-free gas at saturation

For gases the mole ratio is also the partial pressure ratio.

$$n_v = p_v/(P - p_v)$$

$$n_s = P_v/(P - P_v), \text{ where } P \text{ is the total pressure}$$

$$Y_P = [\{p_v/(P - p_v)\}/\{(P_v/P - P_v)\}] \times 100$$

$$= \{(p_v/P_v) \times (P - p_v)/(P - P_v)\} \times 100$$

$$= Y_r \times (P - p_v)/(P - P_v)$$

Percentage saturation is always smaller than relative saturation. The composition of a partially saturated gas–vapour mixture is fixed if the relative or percentage saturation, temperature and pressure are specified.

ILLUSTRATION

5BB A mixture of ether vapour and nitrogen contains 50% ether by volume. Calculate the relative saturation and percentage saturation of the mixture at 295 K and a pressure of 750 mm. Vapour pressure of ether at 295 K is 440 mm Hg.

Solution

For gases, volume percentage, pressure percentage and mole percentage are one and the same. The mole percentage divided by 100 gives the mole fraction.

Partial pressure of ether (p_v) = Total pressure × mole fraction

$= 750 \times 0.5 = 375$ mm Hg

Vapour pressure of ether at 295 K (P_v) = 440 mm Hg

Relative saturation (Y_r) $= (p_v/P_v) \times 100 = \dfrac{375 \times 100}{440} = 85.23\%$

Amount of ether per mole of ether-free nitrogen at the existing conditions (n_v) $= \dfrac{0.5}{0.5} = 1.0$ mol

Amount of ether per mole of ether-free nitrogen at saturation (n_s) $= P_v/p_v = \dfrac{440}{375} = 1.419$ mol

Percentage saturation (Y_p) $= (n_v/n_s) \times 100 = \dfrac{1 \times 100}{1.419} = 70.45\%$

$$Y_p = Y_r \times \frac{(750-440)}{(750-375)}$$

$$= 85.23 \times \frac{(750-440)}{(750-375)} = 70.46\%$$

This proves that Y_p is always less than Y_r.

DEW POINT

If an unsaturated mixture of vapour and gas is cooled, keeping the pressure constant, the partial pressure of vapour will remain the same. The temperature of the mixture will gradually fall till a temperature is attained at which the equilibrium vapour pressure of pure liquid is equal to the existing partial pressure of the vapour in the mixture. The mixture will then be saturated and any further cooling will result in condensation. The temperature at which this takes place is known as the *dew point* (DP) of the mixture.

In other words, at the dew point,

Equilibrium vapour pressure of liquid = partial pressure of vapour in the mixture

The vapour content of a vapour–gas mixture can be predicted or calculated from dew point data.

ILLUSTRATIONS

5CC It is proposed to recover hexane, which is used as a solvent in an extraction process, by evaporation into a stream of nitrogen. Nitrogen enters the evaporator containing hexane at a temperature of 303 K, such that its dew point is 278 K. It leaves at a temperature of 298 K with a dew point of 293 K. The barometric pressure is constant at 750 mm. (i) Calculate the vapour concentrations of the gases entering and leaving the evaporator expressed in moles of vapour per mole of vapour-free gas. (ii) Calculate the number of moles of acetone evaporated per mole of vapour-free gas passing through the evaporator. (iii) Calculate the weight of hexane evaporated per 1000 m^3 of gases entering the evaporator. (iv) Calculate the volume of gases leaving the evaporator per 1000 m^3 of entering gases. Vapour pressure of hexane is 60 mm Hg at 278 K and 135 mm Hg at 293 K.

Solution

(i) *Composition of entering gas*

The gas entering the evaporator has a dew point of 278 K. Hence, the partial pressure of hexane vapour in the entering gases is equal to the vapour pressure of hexane at 278 K.

For gases, partial pressure ratio is also equal to mole ratio.

Partial pressure of hexane at 278 K = 60 mm

Partial pressure of nitrogen = total pressure − partial pressure of hexane

$= 750 - 60 = 690$ mm

Amount of hexane per mole of nitrogen $= \dfrac{60}{690} = 0.087$ mol

(ii) *Composition of leaving gas*

The gas leaving the evaporator has a dew point of 293 K. Hence, the partial pressure of hexane vapour in the leaving gases is equal to the vapour pressure of hexane at 293 K.

Partial pressure of hexane at 293 K = 135 mm

Partial pressure of nitrogen = total pressure − partial pressure of hexane

$= 750 - 135 = 615$ mm

Amount of hexane per mole of nitrogen $= \dfrac{135}{615} = 0.220$ mol

Amount of hexane evaporated per mole of nitrogen = amount of hexane in the leaving gases − amount of hexane in the entering gases

$= 0.220 - 0.087 = 0.133$ mol

(iii) Amount of gases in 1000 m^3 of gas at the entering conditions $= \dfrac{PV}{RT} = \dfrac{750 \times 1000}{760 \times 0.082\,05 \times 303} = 39.70$ kmol

At the inlet, 1 mol of nitrogen is accompanied by 0.087 mol of hexane. For 1 mol of nitrogen or 1.087 mol of entering gases, 0.133 mol of hexane is evaporated.

Amount of hexane evaporated per 1000 m^3 of entering gases $= 0.133 \times \dfrac{39.70}{1.087} = 4.86$ kmol

Molecular weight of hexane is 86.

Weight of hexane evaporated $= 4.86 \times 86 = 417.96$ kg

(iv) Amount of gas leaving = amount of entering gases + amount of hexane evaporated

$= 39.70 + 4.86 = 44.56$ kmol

Volume of leaving gases at 298 K and 750 mm pressure $= \dfrac{nRT}{P} = \dfrac{44.56 \times 0.082\,05 \times 298 \times 760}{750} = 1\,104.06$ m^3

5DD A solvent recovery system delivers a gas saturated with benzene vapour which analyses on benzene-free basis as follows: 15% carbon dioxide, 4% oxygen and the rest nitrogen. This gas is at 298 K and 750 mm Hg. It is compressed to 5 atm and cooled to 298 K after compression. How many kilograms of benzene are condensed by this process per 1000 m^3 of the original mixture? Vapour pressure of benzene at 298 K is 75 mm Hg.

Solution

Basis: 1000 m^3 of original mixture

Amount of gas in original mixture $= \dfrac{PV}{RT} = \dfrac{1 \times 1000}{0.082\,05 \times 298} = 40.90$ kmol

The gases are saturated with benzene at 298 K. So the partial pressure of benzene is equal to the vapour pressure at 298 K.

Mole fraction of benzene $= \dfrac{\text{partial pressure}}{\text{total pressure}} = \dfrac{75}{750} = 0.1$

Amount of benzene in the original mixture $= 40.90 \times 0.1 = 4.09$ kmol

Amount of benzene-free gas $= 40.90 \times 0.9 = 36.81$ kmol

The gases are compressed to 5 atm (3 800 mm Hg) and after compression, cooled to the original temperature, 298 K. So the partial pressure of benzene in the final mixture also is 75 mm Hg.

Mole fraction of benzene in the final mixture $= \dfrac{75}{3\,800} = 0.019\,7$

Mole fraction of benzene-free gases in the final mixture $= 1 - 0.019\,7 = 0.980\,3$

Amount of benzene-free gas $= 36.81$ kmol

Amount of benzene in the final mixture $= 36.81 \times \dfrac{0.019\,7}{0.980\,3} = 0.734$ kmol

Amount of benzene condensed $=$ benzene in the original mixture $-$ benzene in the final mixture

$= 4.09 - 0.734 = 3.356 \text{ kmol} = 261.77 \text{ kg}$

5EE What is the minimum volume of dry air in cubic metres at 293 K and 740 mm of Hg necessary to evaporate 13.8 kg of ethanol, if the total pressure remains constant. Assume that the air blown over the alcohol to evaporate it leaves at 740 mm Hg. Vapour pressure of alcohol at 293 K is 45 mm Hg.

Solution

Basis: 13.8 kg of ethanol

Amount of ethanol $= \dfrac{13.8}{46} = 0.3$ kmol

Minimum amount of air is required only at saturation, when partial pressure of ethanol in the mixture is equal to its vapour pressure.

Partial pressure of ethanol = 45 mm Hg

Partial pressure of air = total pressure – partial pressure of ethanol

$$= 740 - 45 = 695 \text{ mm Hg}$$

For gases, partial pressure ratio is also equal to mole ratio.

$$\frac{45}{695} = \frac{\text{number of kilomoles of vapour}}{\text{number of kilomoles of air}} = \frac{0.3}{\text{kmol of air}}$$

Amount of air = 4.633 kmol

Volume of air at 293 K and 740 mm Hg

$$= \frac{nRT}{P} = \frac{4.633 \times 0.082\,05 \times 293 \times 760}{740} = 14.39 \text{ m}^3$$

HUMIDIFICATION

Humidification is a unit operation involving air–water contact or air–water vapour mixtures.

Dry bulb temperature (DBT)

The dry bulb temperature is the temperature of the air–water vapour system.

Wet bulb temperature (WBT)

The equiiibrium temperature attained by a liquid which is vapourising into a gas is termed as wet bulb temperature, and it is always less than the actual dry bulb temperature of the gas into which evaporation is taking place.

Absolute humidity (Y)

The absolute humidity is the number of kilograms of water vapour present in one kilogram of dry air (DA).

Molal humidity (Y_m)

The molal humidity is the number of moles of water vapour present is one mole of dry air.

$$Y_m = \frac{\text{moles of water vapour}}{\text{moles of air}} = \frac{\text{partial pressure of water vapour}}{\text{partial pressure of air}}$$

$$= \frac{\text{moles of water vapour} \times 18}{\text{moles of air} \times 29}$$

$$Y = \frac{\text{weight of water vapour}}{\text{weight of dry air}}$$

 So, $Y = 0.620\,7\, Y_m$

Saturation humidity (Y_s)

The saturation humidity is the absolute humidity at saturation.

Percentage humidity (Y_p)

The percentage humidity is the percentage ratio of absolute humidity to absolute humidity at saturation.

$$Y_p = \frac{Y}{Y_{\text{at saturation}}} \times 100$$

Relative humidity (Y_r)

The relative humidity is the percentage ratio of partial pressure of water vapour to equilibrium vapour pressure at that temperature.

$$Y_r = \frac{\text{partial pressure}}{\text{vapour pressure}} \times 100$$

Humidity volume (V_y)

The humidity volume is the volume in cubic metres per kilogram of dry air, occupied by one kilogram of dry air, and the accompanying moisture at dry bulb temperature (DBT) and atmospheric pressure.

$$V_y = \left(\frac{1}{29} + \frac{Y}{18}\right) \times 22.4 \times \frac{\text{DBT}}{273}, \text{ where DBT is in Kelvin}$$

Humid heat (C_y)

The humid heat is the heat capacity of one kilogram of dry air and the accompanying moisture.
$C_y = 0.24 + 0.45\,Y$ kcal per kg dry air °C or $1 + 1.84\,Y$ kJ per kg dry air K
where Y is the absolute humidity.

HUMIDITY CHART

The humidity chart is a convenient diagram showing the properties of mixtures of air and water vapour.

Dry bulb temperature (T_{DB}) is plotted as the abscissa and absolute humidity (Y) as the ordinate (Fig. 5.11). Any point on the chart represents a definite mixture of air and water. The curved line gives the humidity of saturated air as a function of air temperature (Y_s vs T). It is known as the *saturation humidity line*. The coordinates of the points on this curve are obtained from vapour pressure–temperature data for water using the equation

$$\frac{\text{partial pressure of water vapour}}{\text{partial pressure of dry air}} \times \frac{18}{29}$$

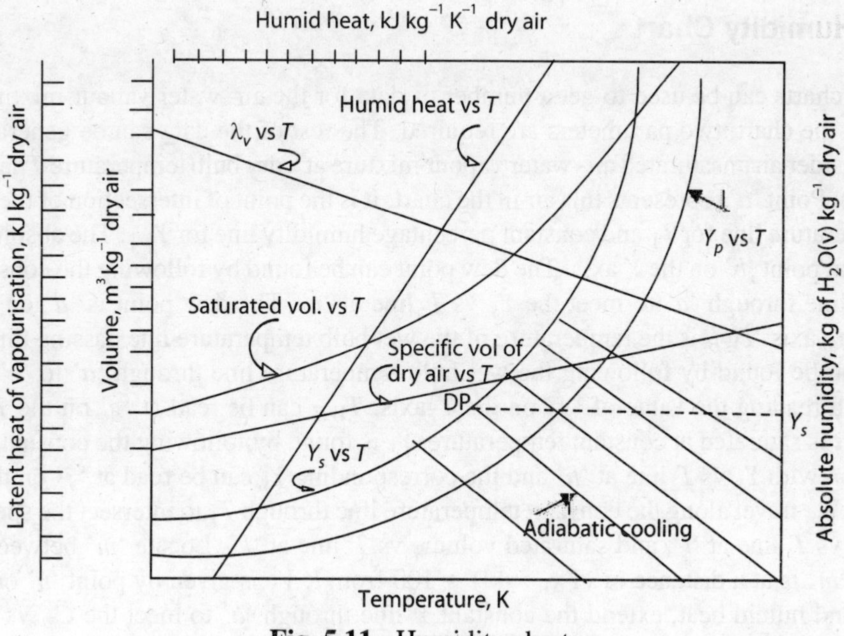

Fig. 5.11 Humidity chart

Any point above and to the left of the saturated line represents a mixture of saturated air and liquid. Any point below the saturated line represents under-saturated air. Any point on the temperature axis represents dry air. Curved lines between the saturation humidity line and the temperature axis marked in % represent mixtures of air and water vapour of definite percentage humidity. They have been drawn based on

$$Y_p = \frac{Y}{Y_{\text{at saturation}}} \times 100$$

Linear interpolation between the saturated line and the temperature axis can be used to locate lines of any given constant Y_p. Slanting lines running down and to the right of the saturated line are wet bulb temperature (T_{WB}) lines. The ends of these lines are identified by the corresponding T_{WB}.

Plots of specific volume of dry air and saturated volume are also available. In both volume is plotted against temperature, and volume can be read on the scale at the left. Coordinates of points on these lines are calculated using the equation

$$V_y = \frac{22.4 \times T}{273} \times \left(\frac{1}{29} + \frac{Y}{18}\right)$$

For dry air, $Y = 0$ and for saturated air, $Y = Y_S$. Linear interpolation between the two lines based on Y_p gives humid volume of unsaturated air. A plot of C_y vs Y is also available. This line is a plot of equation $C_y = C_{p,\text{air}} + YC_{P,\text{water}(v)}$. Some charts include $\lambda_v - T$ plot also, based on $\lambda_v - T$ data. The scale for λ_v is at the extreme left.

Use of Humidity Chart

Humidity charts can be used to get a number of data for the air–water vapour mixture. To locate a point on the chart, two parameters are required. The rest of the data can be generated from the chart. Consider an unsaturated air–water vapour mixture at a dry bulb temperature T_1 and given Y_{p1} (Fig. 5.12). Point '*a*' represents this air in the chart. It is the point of intersection of the constant dry bulb temperature line for T_1 and constant percentage humidity line for Y_{p1}. The absolute humidity, Y, is read at point '*b*' on the Y axis. The dew point can be found by following the constant absolute humidity line through '*a*' to meet the Y_s vs T line at '*c*'. The dew point is '*d*' on the dry bulb temperature axis. T_{WB} is the temperature of the wet bulb temperature line passing through '*a*'. Y_S at T_{WB} can be found by following the wet bulb temperature line through '*a*' to '*e*' at the Y_s vs T line, and reading the value of '*f*' on the Y-axis. T_{WB} can be read at '*g*' on the T-axis. If the original air is saturated at constant temperature, Y_s is found by following the constant T line to its intersection with Y_s vs T line at '*h*' and the corresponding Y_s can be read at '*j*' on the Y-axis.

To find V_y, travel along the constant temperature line through T_1 to intersect the specific volume of dry air vs T line at '*l*', and saturated volume vs T line at '*k*'. Locate '*m*' between '*k*' and '*l*' such that '*m*' is at a distance of $(Y_{p1} \times \text{kl}) \div 100$ from l. V_y is given by point '*n*' on the volume axis. To find humid heat, extend the constant Y line through '*a*' to meet the C_p vs Y line at '*o*' and C_Y can be read at '*p*' on the C_Y scale.

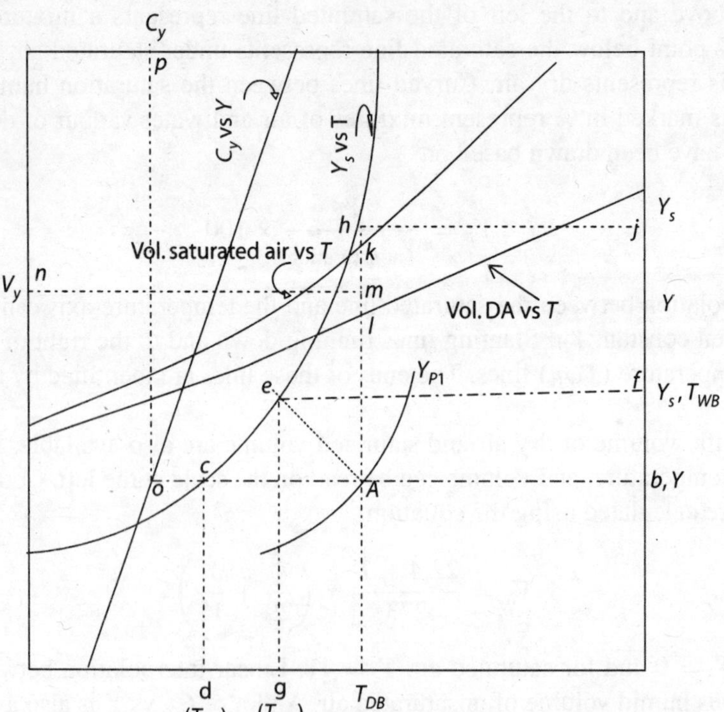

Fig. 5.12 Usage of humidity chart

ILLUSTRATIONS

5FF On a particular day at Chennai the ambient air has the following conditions: Pressure = 752 mm Hg; dry bulb temperature = 306 K; dew point = 295 K. Find the absolute humidity by calculation. Vapour pressure of water at 295 K is 19.8 mm Hg.

Solution

The dew point is a measure of humidity. The partial pressure at dry bulb temperature is equal to vapour pressure at dew point.

Partial pressure of vapour (p_v) = 19.8 mm

Partial pressure of vapour-free gas (p_{vfg}) = total pressure − partial pressure of vapour

= (752 − 19.8) mm

$$Y_m = p_v/p_{vfg}$$

$$= \frac{19.8}{(752-19.8)} = 0.027 \text{ mol of water per mol of DA}$$

$$Y = 0.6207\, Y_m$$

$$= 0.6207 \times 0.027 = 0.0168 \text{ kg water per kg of DA}$$

5GG Air at 298 K and 70% relative humidity is cooled to 288 K and reheated to 293 K. Find (i) the weight of water removed per unit weight of dry air, (ii) the wet bulb temperature of original and final air, and (iii) the percentage humidity, relative humidity and partial pressure of water vapour in the final air.

Solution

The problem has to be solved making use of the humidity chart.

(i) The point of intersection of the constant dry bulb temperature line for 298 K and constant relative humidity line for $Y_r = 70\%$ gives the absolute humidity, Y, as 0.0140 kg per kg of dry air. This gas is cooled to 288 K.

Y_s at 288 K is to be found by following the constant dry bulb temperature line till it meets the Y_s vs T line. Y_s at 288 K is 0.0106 kg per kg of dry air. Humidity of final air is less than that of initial air.

So water removed during cooling = 0.0140 − 0.0106 = 0.0034 kg per kg of dry air.

(ii) Dew point or wet bulb temperature is the point of intersection of the constant absolute humidity line and the Y_s vs T line. The dew point is to be read on the dry bulb temperature axis.

Wet bulb temperature of original air = 294 K

Wet bulb temperature of final air = 290 K

(iii) Y of final air = 0.0106

The final air is reheated to the original temperature 293 K.
Y_s of final air = 0.0146, as read from humidity chart for temperature 293 K.

We know that $Y_P = \dfrac{Y}{Y_{\text{at saturation}}} \times 100 = 72.6\%$

Y_m of final air $= 0.0106 \div 0.6207 = 0.0171$

Y_m of saturated air at 293 K $= 0.0146 \div 0.6207 = 0.0235$

Y_m is the molar ratio of water vapour to dry air, which is also equal to the partial pressure ratio of water vapour to dry air.

$0.0171 = p_v \div (760 - p_v)$

$0.0235 = P_v \div (760 - P_v)$

So, p_v or partial pressure of water vapour = 13.22 mm Hg

P_v or vapour pressure of water = 17.54 mm Hg

Relative humidity, $Y_r = p_v \div P_v$ = 75.37%

Partial pressure of water vapour in the final air is 13.22 mm Hg.

5HH An air–water vapour mixture of 740 mm Hg and 298 K has a dew point of 291 K. It is desired to remove 70% of water by cooling to 281 K and compressing. Calculate the necessary pressure required. Vapour pressure of water is 15.46 mm Hg at 291 K and 8.03 mm Hg at 281 K.

Solution

The dew point is a measure of humidity. The vapour pressure at the dew point is equal to the partial pressure at dry bulb temperature.

Initial water content of the mixture $= \dfrac{15.46}{(740 - 15.46)} = 0.0213$ mol per mole of DA

70% of moisture is removed by cooling to 281 K and compressing.

Let the final pressure be P.

$$P = 0.0213 \times \frac{30}{100} = 0.00639 \text{ mol per mole of DA.}$$

The final air is saturated at 281 K. So, partial pressure of moisture in the final air will be equal to vapour pressure at 281 K. The molar ratio is equal to the partial pressure ratio.

$$\frac{0.00639}{1} = \frac{8.03}{(P - 8.03)}$$

Pressure required $(P) = 1264.68$ mm Hg

5II Air at a temperature of 303 K and pressure of 750 mm Hg has a relative humidity of 80%. Calculate (i) the absolute humidity of the air, (ii) the absolute humidity of this air if the temperature is reduced to 288 K and pressure is increased to 2 atm condensing out some water, and (iii) the weight of water condensed during cooling and compression. Vapour pressure of water is 31.8 mm Hg and 12.75 mm Hg at 303 K and 288 K respectively.

Solution

Figure 5.13 represents the data provided in the form of a block diagram.

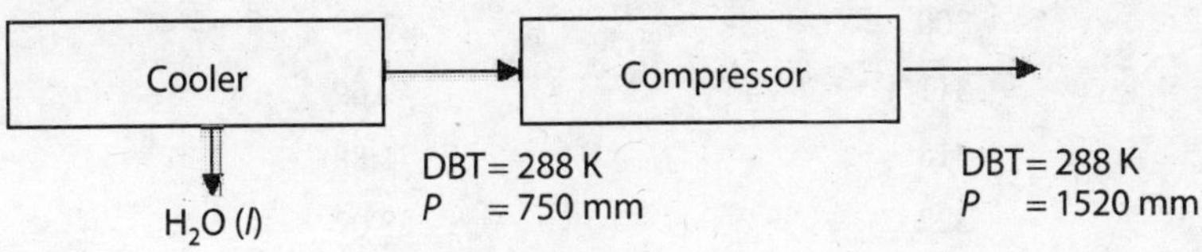

Fig. 5.13 Air conditioning operation

(i) At the inlet, the dry bulb temperature is 303 K, the pressure (P) is 750 mm Hg, and the relative humidity is 80%. At 303 K the vapour pressure of water is 31.8 mm Hg.

Relative humidity is the percentage ratio of partial pressure of water vapour (p_v) to vapour pressure (P_v) at that temperature.

$$Y_r = (p_v \div P_v) \times 100 = 80$$

$$\text{So } p_v = 31.8 \times 0.8 = 25.44 \text{ mm Hg}$$

$$\text{Then } Y_m = p_v \div (P - p_v) = \frac{25.44}{(750 - 25.44)} = 0.0351 \text{ mol per mol of DA}$$

$$Y = Y_m \times 0.6207 = 0.02176 \text{ kg per kg of DA}$$

(ii) If the temperature is reduced to 288 K and the pressure is increased to 2 atm, condensing out some water, air becomes saturated at 288 K. The partial pressure of water vapour becomes equal to the vapour pressure of water at 288 K

$$Y_m \text{ at the outlet} = p_v \div (P - p_v) = \frac{12.75}{(1\,520 - 12.75)} = 0.008\,46 \text{ mol per mol of DA}$$

$$Y = Y_m \times 0.620\,7 = 0.005\,24 \text{ kg per kg of DA}$$

(iii) $$\text{Weight of water removed in the process} = \text{moisture content of air at outlet} - \text{moisture content of air at inlet}$$

$$= 0.021\,76 - 0.005\,24$$

$$= 0.016\,52 \text{ kg per kg of DA}$$

5JJ Air at a temperature of 311 K and atmospheric pressure has a dew point of 300 K. (i) Estimate the molal humidity, the molal humidity at saturation and the percentage humidity. The air of part (i) is passed into an evaporator from where it emerges at a temperature of 522 K with a dew point of 518 K. (ii) Estimate the percentage humidity of the air leaving the evaporator, and (iii) calculate the weight of water evaporated per 1000 m^3 of air entering.

Temperature (K)	Vapour pressure of water (mm of Hg)
300	27.74
311	49.69
318	71.88
322	88.02

Solution

The dew point is a measure of moisture content. The vapour pressure of water at the dew point is equal to the partial pressure of water vapour in the mixture.

(i) *Entering air*

$$Y_m = \frac{26.74}{(760 - 26.74)} = 0.036\,5 \text{ mol per mole of DA}$$

$$Y_m \text{ at saturation} = \frac{49.69}{(760 - 49.69)} = 0.07 \text{ mol per mole of DA}$$

$$\text{Percentage saturation} = \frac{Y_m}{Y_{m\text{ (at saturation)}}} \times 100 = 52.14\%$$

(ii) *Leaving air*

$$Y_m = \frac{71.88}{(760 - 71.88)} = 0.104\,5 \text{ mol per mole of DA}$$

$$Y_m \text{ at saturation} = \frac{88.02}{(760 - 88.02)} = 0.131\,0 \text{ mol per mole of DA}$$

$$\text{Percentage saturation} = \frac{Y_m}{Y_m \text{ (at saturation)}} \times 100 = 79.77\%$$

(iii) Amount of entering air $= \dfrac{PV}{RT} = \dfrac{1 \times 1000}{0.082\,05 \times 311} = 39.188$

Water vapour picked up per kmol of DA + 0.036 5 kmol of water vapour or 1.036 5 kmol of entering air = moisture in leaving air − moisture in entering air

$$= 0.104\,5 - 0.036\,5 = 0.068 \text{ kmol}$$

$$\text{Total water evaporated} = \frac{0.068}{1.036\,5} \times 39.188 = 2.571 \text{ kmol}$$

$$= 46.28 \text{ kg}$$

5KK An air conditioning system takes outside air at 303 K and 90% humidity, cools it to 288 K and reheats the cooled air to 298 K. Calculate (i) the weight of water vapour removed per kilogram of dry air in the cooling plant, (ii) the wet bulb temperature of the origin and conditioned air, and (iii) the percentage humidity, the relative humidity and the partial pressure of water vapour in the conditioned air.

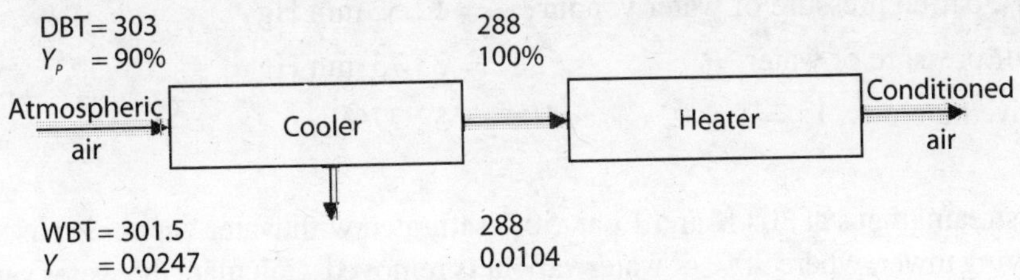

Fig. 5.14 Air conditioning operation

Solution

Figure 5.14 represents the data provided in the form of a block diagram. The problem has to be solved making use of the humidity chart.

(i) The point of intersection of the constant dry bulb temperature line for 303 K and constant percentage humidity line for $Y_p = 90\%$ gives the absolute humidity, Y, as 0.024 7 kg per kg of dry air. This gas is cooled to 298 K. Y_s at 298 K is to be found by following the constant dry bulb temperature line till it meets the Y_s vs T line. Y_s at 288 K is 0.0104 kg per kg of dry air. Humidity of final air is less than that of initial air.

So water removed during cooling $= 0.0247 - 0.0104 = 0.0143$ kg per kg of dry air

(ii) Dew point or wet bulb temperature is the point of intersection of the constant absolute humidity line and the Y_s vs T line. The dew point is to be read on the dry bulb temperature axis.

Wet bulb temperature of original air $= 301.5$ K

Wet bulb temperature of final air $= 291.5$ K

(iii) Y of final air $= 0.0104$

The final air is reheated to 298 K.

Y_s of final air is 0.02, as read from the humidity chart for temperature 298 K.

$$\text{We know that, } Y_p = \frac{Y}{Y_{\text{at saturation}}} \times 100 = 52\%$$

Y_m of final air $= 0.0104 \div 0.6207 = 0.0167$

Y_m of saturated air at 298 K $= 0.02 \div 0.6207 = 0.0322$

Y_m is the molar ratio of water vapour to dry air , which is also equal to partial pressure ratio of water vapour to dry air.

$$0.0167 = p_v \div (760 - p_v)$$

$$0.0322 = P_v \div (760 - P_v)$$

So, p_v, partial pressure of water vapour $= 12.52$ mm Hg

Vapour pressure of water, P_v $= 23.73$ mm Hg

Relative humidity, $Y_r = (p_v \div P_v) \times 100 = 52.77\%$

5LL A stream of gas at 304 K and 1 bar, 50% saturated with water vapour is passed through a drying tower, where 90% of water vapour is removed. Calculate the water vapour content of entering gas and leaving gas, weight of water removed per 1000 m^3 of entering gas and volume of leaving gas per 1000 m^3 of entering gas, if the gas leaves at 300 K and 0.98 bar. Vapour pressure of water at 304 K is 0.36 bar.

Solution

Basis: 1000 m^3 of entering gas

At saturation, the partial pressure is equal to the vapour pressure at that temperature. The molar ratio is also the partial pressure ratio.

Saturation humidity at inlet conditions $= \dfrac{0.36}{1 - 0.36} = 0.562\,5$ mol per mol DA

Entering gas is 50% saturated.

Molar humidity of incoming air $= 0.562\,5 \times \dfrac{50}{100} = 0.281\,3$ mol per mole of DA

90% of water vapour is removed and the outgoing gas contains only 10% of initial water vapour content.

Molar humidity of outgoing air $= 0.281\,3 \times \dfrac{10}{100} = 0.028\,1$ mol per mole of DA

Amount of incoming air $= \dfrac{PV}{RT} = \dfrac{1 \times 1000}{1.013 \times 0.082\,05 \times 304} = 39.57$ kmol

1 mol of entering dry air contains 0.2 813 mol of moisture.

Amount of dry air $= \dfrac{39.57}{(1 + 0.281\,3)} = 30.88$ kmol

1 mol of leaving dry air contains 0.028 13 mol of moisture.

Amount of outgoing gas $= 30.88 \times 1.028\,13 = 31.75$ kmol

Water removed = entering air – leaving air

$= 39.57 - 31.75 = 7.82$ kmol $= 140.76$ kg

Volume of outgoing gas $= \dfrac{nRT}{P} = \dfrac{31.75 \times 0.082\,05 \times 300 \times 1.013}{0.98} = 807.84\ \text{m}^3$

5MM Hot air is being used to dry a wet material. The hot air enters the drier at a total pressure of 770 mm Hg and temperature of 433 K. The partial pressure of water vapour in the incoming air is 25 mm Hg. At the exit of the drier, the temperature is 383 K and the pressure is 760 mm Hg. In the outgoing air, the partial pressure of water vapour is 100 mm Hg. Calculate the absolute humidity of incoming air and outgoing air. Also calculate the weight of moisture picked up per 1000 m^3 of entering air.

Solution

Basis: 1000 m^3 entering gases

Incoming air

Molar humidity of incoming air $= \dfrac{\text{partial pressure of moisture}}{\text{partial pressure of dry air}}$

$= \dfrac{25}{(770 - 25)} = 0.033\,6$ mol per mole of DA

Absolute humidity $= \text{molal humidity} \times 0.6207$

$= 0.0336 \times 0.6207 = 0.0208$ kg per kg of DA

Outgoing air

Molar humidity of incoming air $= \dfrac{\text{partial pressure of moisture}}{\text{partial pressure of dry air}}$

$= \dfrac{100}{(760 - 100)} = 0.1515$ mol per mole of DA

Absolute humidity $= \text{molal humidity} \times 0.6207$

$= 0.1515 \times 0.6207 = 0.094$ kg per kg of DA

Moisture picked up per kilomole of DA $=$ moisture at outlet – moisture at inlet

$= 0.1515 - 0.0336 = 0.1179$ kmol

Amount of air entering $= \dfrac{PV}{RT} = \dfrac{1000 \times 770}{0.08205 \times 433 \times 760} = 28.52$ kmol

At the inlet 1 mol of dry air is accompanied by 0.0336 mol moisture. 1 mol of dry air or 1.0336 mol of entering moist air picks up 0.1179 mol moisture.

Total moisture picked up $= \dfrac{0.1179}{1.0336} \times 28.52 = 3.253$ kmol $= 58.55$ kg

Problems for Practice

5.1 In one hour, 5000 kg of an aqueous solution containing 10% sodium chloride has to be concentrated to 50% sodium chloride solution in a double effect evaporator. The feed solution is fed to the first effect where it is concentrated to 25% sodium chloride solution, which in turn is fed to the second effect to produce the desired product. Compute (i) the amount of water evaporated in each effect per hour, (ii) the water evaporating capacity of the plant in tonnes per day, and (iii) the production of 50% sodium chloride solution per shift of 8 hours.

5.2 A mixture of ammonia and air at 750 mm Hg pressure and 303 K contains 7.2% ammonia by volume. The gas is passed at the rate of 200 $m^3\,h^{-1}$ through an absorption tower where only ammonia is absorbed. The gases leave the tower at 720 mm Hg pressure and 293 K with 0.1% ammonia. Compute the weight of ammonia absorbed and the volume of the leaving gases.

5.3 Absorption of sulphur dioxide is carried out in a packed tower. Gas containing 14.8% of sulphur dioxide and the rest inert gases enters the tower, while the leaving gases contain 1%

sulphur dioxide. Water flows at the rate of 1000 L per min. The tower handles 7 380 $m^3\,h^{-1}$ of gas at 303 K and 1 bar. Find the sulphur dioxide concentration in the effluent from the tower.

5.4 A mixture of ammonia and air at 720 mm Hg pressure and 313 K contains 6.3% ammonia by volume. The gas is passed at the rate of 100 $m^3\,h^{-1}$ through an absorption tower where only ammonia is absorbed. The gases leave the tower at 715 mm Hg pressure and 303 K with 0.09% ammonia. Compute the weight of ammonia absorbed and the volume of the leaving gases.

5.5 A distillation column separates a mixture of ethanol and water into high purity alcohol and waste water. The feed is a mixture of 20mol% ethanol and 80mol% water. The distillate is to contain 85mol% ethanol and the residue is to have 3mol% ethanol. Calculate (i) the quantity of distillate and residue produced per 1000 kg of feed, and (ii) the percentage of alcohol in the feed that has been recovered as distillate.

5.6 1000 kg of a mixture of 15% methyl alcohol and the rest water is fed to a distillation column. The product leaving the top of the column contains 85% alcohol. 5% of the alcohol fed to the column goes to the residue. Calculate the product obtained and the composition of the residue. All percentages are by weight.

5.7 A refinery separates 50 000 kg h^{-1} of a mixture containing 35% propane and 65% butane. 100 kg of butane appear in the distillate (the remainder being propane), and 65 kg of propane in the residue (the remainder being butane). Give the analysis of the distillate and the residue.

5.8 1000 kg of a solution of citric acid and water contains 5% acid and 95% water. Citric acid is removed from water using benzene as the extraction solvent. Benzene and water are immiscible. When an aqueous solution of acid is thoroughly mixed with pure benzene and allowed to settle, two layers of liquid are formed. The benzene layer contains 90% benzene and 10% citric acid; the water layer contains 1% benzene, 0.5% acid and the rest water. Calculate the weight of benzene used in this operation.

5.9 Soya bean flakes contain 15% oil. A lot of 500 kg of these flakes is extracted with 500 kg of a solvent containing 1% oil. The flakes and solvent are mixed together thoroughly and the solution is drained off. The flakes are still wet with the solution. Oil contained in this solution is 3 kg. The composition of the solution retained on the flakes is the same as that of the solution drained off. Compute the amount of solution obtained and its composition.

5.10 Dolomite ore is a mixture of calcium and magnesium carbonates. It is leached with concentrated sulphuric acid to recover magnesium as magnesium sulphate. Dolomite contains 25% calcium, 15% magnesium and 5% silica. After leaching, the soluble matter is filtered off and the solids are washed. The washed solids are analysed and found to contain 2% silica, 0.5% magnesium and 1% aluminum. Compute the percentage of magnesium extracted by acid leaching.

5.11 The bark of a certain tree contains 6% moisture, 10.8% tannin and 8.2% soluble non-tannin matter, the balance being insoluble lignin. After extraction by the counter current Shank's process, the waste bark contains 0.95% tannin and 0.22% soluble non-tannin on a dry basis. If raw tannin fetches Rs 50 per kg, how much of money is lost per tonne of original bark as a result of imperfect extraction?

5.12 A saturated solution of magnesium sulphate at 353 K is cooled to 303 K. During cooling 4% of original water is lost. Calculate the quantity of original solution needed to crystallise 1000 kg of $MgSO_4.7H_2O$. The solubility in terms of weight of magnesium sulphate per 100 gm of water is 40.8 and 64.2 at 303 K and 353 K respectively.

5.13 In a crystallisation process 2 500 kg of $Na_2SO_4.10H_2O$ crystals are obtained. The mother liquor leaving the process contains 20% Na_2SO_4 and the feed contains 36% Na_2SO_4. During the process 25% of original water is lost by evaporation. Find the weights of feed required, mother liquor leaving and water evaporated.

5.14 A crystalliser is charged with 9 774 kg h^{-1} of an aqueous solution having 30% Na_2SO_4. The solution is cooled and 5% of the feed weight is lost as water of evaporation. After the $Na_2SO_4.10H_2O$ crystals were obtained, the mother liquor was found to contain 18% Na_2SO_4. Find the weight of crystals obtained, mother liquor and water loss.

5.15 5 000 kg of potassium chloride are present in a saturated solution at 353 K. The solution is cooled to 293 K in an open tank. The solubilities are 55 and 25 parts per 100 parts of water at 353 K and 293 K respectively. Assuming that 3% of the solution evaporates as water, find the weight of mother liquor, water evaporated and crystals formed.

5.16 9 581 kg of a 30% solution of sodium carbonate is slowly cooled to 293 K. During cooling 10% of the original water is lost by evaporation. The crystals have ten molecules of water of hydration, and the mother liquor carries 21.6 kg of salt per 100 kg of water. Find the weight of the streams leaving the crystalliser.

5.17 A solution of potassium chloride in water is saturated at a temperature of 288 K. Calculate the weight of potassium chloride that can be dissolved by 100 kg of this solution, if it is heated to a temperature of 338 K. Solubility of potassium chloride at 288 K is 1.62 kmol per 100 kg water; solubility of potassium chloride at 338 K is 6.37 kmol per 100 kg water.

5.18 300 kg of a saturated solution of silver nitrate at 373 K are cooled to 293 K and the crystals are filtered out. Wet filter cake contains 80% solid crystals and 20% mother liquor. The wet cake is dried to remove water by evaporation. Calculate the fraction of silver nitrate in the feed stream removed eventually as dry crystals and the amount of water evaporated during the drying operation. Solubility of silver nitrate in water: 952 kg of silver nitrate per 100 kg of water at 373 K, and 222 kg of silver nitrate per 100 kg of water at 293 K.

5.19 Air at a temperature of 293 K and pressure 0.987×10^5 N m^{-2} has a relative humidity of 80%. Calculate (i) the molal humidity, (ii) the molal humidity if the temperature is reduced to 283 K and pressure increased to 2.38×10^5N m^{-2}, condensing some of the water, (iii) the weight of condensed water from 29.3 m^3 of original wet air in cooling and compressing to the conditions as mentioned in (ii), and (iv) the final weight of wet air of part (iii). Vapour pressures are 0.023×10^5 and $0.001\,2 \times 10^5$ N m^{-2} at 293 K and 283 K respectively.

5.20 In a solvent recovery unit, 50 kg of acetone is to be recovered by a stream of nitrogen. Nitrogen is admitted at 308 K and 75 cm Hg pressure. Partial pressure of acetone in the incoming nitrogen is 10 mm Hg. The gases leave at 303 K and 740 mm Hg pressure with 90% saturated vapour. Find the volume of nitrogen required for this duty and the volume of the gases leaving the unit. Vapour pressure of acetone at 303 K is 270 mm Hg.

5.21 Air entering a drier has a dew point of 280 K, while that of leaving air is 290 K. If 2 500 kg of wet air is forced through the drier, how many kilograms of water are removed? Assume that the total pressure remains the same at 100 kN m^{-2}. Vapour pressure of water at 280 K and 290 K are 1 kN m^{-2} and 2 kN m^{-2} respectively.

5.22 What is the minimum volume of dry air in cubic metres required at 298 K and 750 mm Hg required to evaporate 11.7 kg of benzene if the total pressure remains constant? Assume that the air blown over the benzene to evaporate leaves at 750 mm Hg. Vapour pressure of benzene at 298 K is 75 mm Hg.

5.23 Air at 333 K and 760 Torr with a percentage humidity of 10% is fed to the drier at the rate of 1000 m^3 h^{-1}. In the drier water is removed from a wet solid material at the rate of 20 kg h^{-1}. Air leaves the drier at 313 K and 740 Torr. Compute the flow rate of air leaving the drier. Vapour pressure of water at 333 K is 150 mm Hg.

5.24 A mixture of dry flue gases and acetone at a pressure of 750 mm Hg has a dew point of 298 K. It is proposed to condense 90% of the acetone by cooling to 278 K and compressing. Calculate the final pressure of the mixture. The vapour pressure of acetone is 230 and 90 mm Hg at 298 K and 278 K respectively.

5.25 The dry bulb temperature and the dew point of ambient air were found to be 302 K and 291 K respectively. The barometer reads 100 kPa. Compute (i) the absolute molal humidity, (ii) the absolute humidity, (iii) the percentage relative humidity, (iv) the percentage saturation, (v) the humid heat, and (vi) the humid volume. Vapour pressure of water at 291 K and 302 K is 2.062 4 and 4.004 kPa respectively.

5.26 At 320 K, benzene exerts a vapour pressure of 200 mm Hg. If a mixture containing 20mol% benzene in nitrogen at 320 K and 800 mm Hg pressure is isothermally compressed until the final gaseous volume becomes 20% of the original volume, what weight of benzene is condensed per 1000 m^3 of original mixture?

5.27 Dry nitrogen is bubbled through benzene at 800 mm Hg and 323 K at a rate of 100 m^3 per minute. Nitrogen saturated with benzene vapour leaves at 740 mm Hg and 313 K at a rate of 138.5 m^3 per minute. Calculate the vapour pressure of benzene at 313 K and the composition of the final mixture in kilograms of benzene vapour per kilogram of nitrogen.

6 Unsteady State Material Balance

In a steady state process, conditions at any one place in the system are constant and do not change with respect to time. Conditions will be different at different points in the system but remain constant at each particular point. In other words, the point conditions are invariant with time.

In the case of an unsteady state process, the point conditions vary with time. These are processes in which quantities or operating conditions within the system change with time. In other words, an unsteady state situation will arise whenever the quantity or composition of one of the feed streams is a function of time. This situation normally occurs during:

1. All batch operations like distillation, crystallisation and leaching;
2. All semi-continuous or regenerative operations like adsorption, ion exchange and catalysis;
3. Continuous operations or 'steady state processes' during periods of 'start up', 'shutdown' or 'upset'.

Material balance problems of unsteady state operations are solved by taking into consideration a small time interval and finding out the change in the parameter during that time interval. The solutions involve integral calculus. Mixing in a stirred tank is a typical example of an unsteady state operation.

Mixing in a stirred tank

Let M be the mass of solution in the tank (Fig. 6.1) and x be the mass fraction of solute. The tank is fed with a solution at a mass flow rate of m_i and the mass fraction of solute in the input is x_i. The contents of the tank are thoroughly mixed. The solution flows out of the tank at a mass flow rate of m_o, and the mass fraction of the solute in the output is x_o. Material balance during a short time interval, dt, can be worked out as follows.

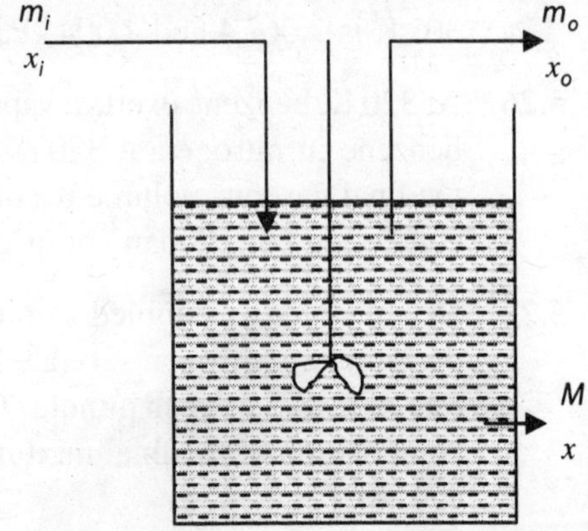

Fig. 6.1 Stirred tank

Input

Mass flow rate of the solution $= m_i$
Mass fraction of the solute $= x_i$

Output

Mass flow rate of the solution $= m_o$
Mass fraction of the solute $= x_o$

In a time interval, dt

Solution input $= m_i dt$
Solution output $= m_o dt$
Accumulation $= m_i dt - m_o dt$
Material balance for the solution: $m_i dt = m_o dt + dM$
Material balance for the solute: $m_i x_i dt = m_o x_o dt + d(Mx)$

$$m_i x_i dt = m_o x_o dt + x dM + M dx$$

Case I If mass in the tank is kept constant, $dM = 0$

$$m_i x_i dt = m_o x_o dt + M dx$$

Case II If the solution is mixed thoroughly so that the outlet composition is same as that in the tank, $x_o = x$

$$m_i x_i dt = m_o x dt + x dM + M dx$$

Case III If the rate of input to the tank equals rate of output from the tank, $m_i = m_o = m$

$$m_i x_i dt = m x_o dt + x dM + M dx$$

Case IV If all the above conditions are to be satisfied, that is, $dM = 0$, $x_o = x$, $m_i = m_o = m$.

$$m x_i dt = m x dt + M dx$$

ILLUSTRATIONS

6A A tank initially contains 500 kg of an aqueous solution of 10% salt. How much water must be fed to the tank to reduce the salt content to 1%? The contents of the tank are thoroughly mixed and the mass in the tank is maintained at 500 kg by keeping the continuous feed and drain streams at equal rates.

Solution

Basis: 1000 kg of dried material

The general equation to be used is,

$$m_i x_i dt = m_o x_o dt + M dx + x dM$$

12 For the present case, $dM = 0$, $x_o = x$ and $m_i = m_o = m$.

As water is fed, $u_i = 0$.

$$0 = mxdt + Mdx$$

$$mdt = \frac{-Mdx}{x}$$

$$m\int_{t=0}^{t=t} dt = -M\int_{x=0.1}^{x=0.01} \frac{dx}{x}$$

$$mt = M[-\ln(0.01/0.1)]$$

$$= 2.303M$$

$$= 2.303 \times 500$$

$$= 1\,151.5\,\text{kg}$$

6B A tank contains 5 m^3 of a 20mass% solution of sodium chloride. Water is added at the rate of 3 m^3 h^{-1} and the solution is withdrawn at the same volumetric rate. Determine the mass of salt remaining after one hour. Assume thorough mixing in the tank and neglect density changes due to concentration. The tank can hold only 5 m^3 of solution. How long will it take for the mass per cent of the solute to become 5?

Solution

The equation to be used is,

$$m_i x_i dt = m_o x_o dt + Mdx + xdM$$

For the present case, $dM = 0$, $x_o = x$ and $m_i = m_o = m$.

As water is fed, $x_i = 0$.

$$0 = mxdt + Mdx$$

$$mdt = \frac{-Mdx}{x}$$

(1) $$m\int_0^1 dt = -M\int_{0.2}^{x} \frac{dx}{x}$$

$$3 = -5\ln(x/0.2)$$

$$= 5[\ln 0.2 - \ln x]$$

$$= 5 \times 2.303[\log 0.2 - \log x]$$

$$x = 0.1097$$

Mass per cent of solute remaining after one hour = 10.97mass%

(2) $m\int_0^t dt = -M\int_{0.2}^{0.05}\frac{dx}{x}$

$$3t = -5\ln 0.25$$

$$t = 2.31\,\text{h}$$

It will take 2.31 hours for the amount of solute to become 5mass%.

6C A certain process requires 100 m^3 of water per hour with a Mg^{++} content of 2 ppm. Mg^{++} content in raw water is reduced to this level by polishing raw water with a specific cation exchange resin. Treated water is fed through a storage tank of capacity 2 000 m^3. The inflow to and outflow from the storage tank are maintained at 100 m^3 h^{-1}. Due to some process lapse in the treatment plant, Mg^{++} content in the treated water increases to 5 ppm. Calculate the time taken for the Mg^{++} content of the water in the storage tank to increase to 3 ppm. Assume thorough mixing in the tank.

Solution

The equation to be used is

$$m_i x_i dt = m_o x_o dt + M dx + x dM$$

For the present case, $dM = 0$, $M = 2\,000$, $m_i = m_o = 100$, $x_o = x$ and $x_i = 5$.

$$100 \times 5 \times dt = 100x dt + 2\,000dx$$

$$dt = \frac{20}{(5-x)}dx$$

when $t = 0$, $x = 2$ and when $t = t$, $x = 3$

$$\int_0^t dt = 20\int_2^3 \frac{dx}{(5-x)}$$

$$t = -20\ln[(5-3)/(5-2)]$$

$$= -20\ln(2/3)$$

$$= 20\ln(3/2)$$

$$= 8.11\,\text{h}$$

6D A storage tank of volume 100 m^3 containing methane has to be flushed by means of nitrogen. If the flow rate of nitrogen is 30 $m^3\ min^{-1}$, determine the time required to reach a gas composition of 5% methane by volume in the tank, when the process is carried out at atmospheric pressure and 303 K.

Solution

$$\text{Nitrogen fed to the tank} = \frac{30 \times 1}{0.082\,05 \times 303} dt$$

$$= 1.207 dt \text{ kmol}$$

Nitrogen fed to the tank mixes with methane and leaves the tank along with it, thereby removing methane from the tank.

Let c kmol m^{-3} be the concentration of N_2 in the tank.

Nitrogen leaving the tank = $30c\,dt$ kmol.

Accumulation of nitrogen in the tank = $100dc$ kmol.

Where dc is the change in concentration during the time interval dt

$$1.207dt - 30cdt = 100dc$$

$$(1.207 - 30c)dt = 100dc$$

$$dt = \frac{100dc}{(1.207 - 30c)}$$

When only 5% of methane is remaining, nitrogen concentration in the outlet from the tank will be $= \dfrac{0.95 \times 1}{0.082\,05 \times 303} = 0.038\,2$ kmol m^{-3}.

When $t = 0$, $c = 0$ and when $t = t$, $c = 0.038\,2$ kmol m^{-3}

$$\int_0^t dt = \int_0^{0.0382} \frac{100dc}{(1.207 - 30c)}$$

$$= \frac{100}{30} \times \ln[1.207/(1.207 - 30 \times 0.038\,2)]$$

$$= 9.95 \text{ min}$$

6E In continuous kraft pulp bleaching, sodium hydroxide is required at a concentration of 10wt% and a flow rate of 1.65 kg s^{-1}. This solution is prepared by introducing 50% sodium hydroxide lye, diluting with water continuously in a 1 900 L tank equipped with an agitator, and withdrawing the 10% solution continuously at the desired rate. During the operation, the inflow of 50% sodium hydroxide lye suddenly fails while the water for dilution continues to enter the tank. Assuming that perfect mixing takes place in the tank and volume in the tank is constant, calculate the time taken for the effluent concentration to fall to 8%. Assume specific gravity of sodium hydroxide in the range 8%–10% to be 1.1.

Solution

For this situation, the equation to be used reduces to

$m_i x_i dt = m_o x_o dt + M dx$, as $dm = 0$

10% sodium hydroxide is made by mixing 20 kg of 50% sodium hydroxide with 80 kg water. When the flow of 50% sodium hydroxide solution is stopped, only 80 kg of water is flowing.

$m_i = 80$; $x_i = 0$; but $m_o = 100$ kg; $x_o = x$

So, $100x dt = -M dx$

$100 dt = -M dx / x M = 1\,900 \times 1.1 = 2\,090$ kg

When $t = 0$, $x = 0.10$ and when $t = t$, $x = 0.08$

$$100 \int_0^t dt = 2\,090 \int_{0.10}^{0.08} \frac{dx}{x}$$

$$100t = 2\,090 \ln(0.10/0.08)$$

$$= 2\,090 \times 0.223\,1$$

$$t = 4.66 \text{ min}$$

6F A small still separates propane and butane at 408 K and initially contains 10 kmol of a mixture, with composition $x = 0.3$, where x is the mole fraction of butane. Additional mixture with $x = 0.3$ is fed at the rate of 5 kmol per hour. The total volume of the liquid in the still is constant and the concentration of vapour from the still (x_D) is related to x_S as follows

$$x_D = \frac{x_S}{1 + x_S}$$

How long will it take for the value of x_S to change from 0.3 to 0.4?. What is the steady sate (equilibrium) value of x_S in the still?

Solution

For this situation, the general equation reduces to

$m_i x_i dt = m_o x_o dt + M dx$

Substituting the values

$5 \times 0.3 \times dt = 5x_D dt + 10 dx_S$

Eliminating x_D in terms of x_S,

$$1.5dt = \frac{5x_S}{1+x_S}dt + 10dx_S$$

$$0.15dt = \frac{0.5x_S}{1+x_S}dt + dx_S$$

$$0.15dt - \frac{0.5x_S}{1+x_S}dt = dx_S$$

$$\left(0.15 - \frac{0.5x_S}{1+x_S}\right)dt = dx_S$$

$$\frac{(0.5 - 0.35x_S)}{1+x_S}dt = dx_S$$

$$dt = \frac{1+x_S}{(0.15+0.35x_S)}dx_S$$

When $t = 0$, $x_S = 0.3$ and when $t = t$, $x_S = 0.4$.

$$\int_0^t dt = \int_{0.3}^{0.4} \frac{1+x_S}{(0.15+0.35x_S)}dx_S$$

$$t = \left[-\frac{0.4}{0.35} - \frac{\ln(0.15 - 0.35 \times 0.4)}{0.35 \times 0.35}\right] - \left[-\frac{0.3}{0.35} - \frac{\ln(0.15 - 0.35 \times 0.3)}{0.35 \times 0.35}\right] = 5.85\,\text{h}$$

Steady state value of x_S is obtained when accumulation is zero, that is, when Input = Output. Composition of input and output does not change with time.

$0.15 = 0.5x_S/(1+x_S)$

$x_S = 0.428\,\text{h}$

Problems for Practice

6.1 A tank of capacity 50 m^3 contains 50 m^3 of a 20mass% solution of sodium chloride. Water is added at the rate of 5 m^3 per hour and is withdrawn at the same volumetric rate. Compute the mass of sodium chloride remaining after one hour. Also determine the time required for reducing sodium chloride content of the solution in the tank to 10mass%. Assume thorough mixing in the tank and neglect density changes due to concentration.

6.2 A storage tank of demineralised water (DM) has a holding capacity of 1 500 m^3 upto an overflow point. DM water having a silica content of 5×10^{-3} ppm is flowing into the tank at the rate of 25 L s^{-1}. The overflow to a high pressure boiler is 25 L s^{-1}. With time, the quality of DM water deteriorates and silica content in the feed DM water increases to 20×10^{-3} ppm. Immediately the inflow of DM water into the tank reduces to 15 L s^{-1} and the outflow is maintained at 25 L s^{-1}. Calculate the time required for the silica content in the tank to increase to 8×10^{-3} ppm.

6.3 A storage tank of DM water has a holding capacity of 1000 m^3 upto an overflow point. DM water having silica content of 5×10^{-3} ppm is flowing into the tank at a rate of 25 L s^{-1}. The overflow to a high pressure boiler is 25 L s^{-1}. With time, the quality of DM water deteriorates and silica content in the feed DM water increases to 20×10^{-3} ppm. Calculate the time required for the silica content in the tank to increase to 10×10^{-3} ppm.

6.4 A tank holds 100 L of a salt solution in which 4 kg of salt is dissolved. Water runs into the tank at the rate of 5 L per minute and salt solution overflows at the same rate. If the mixing in the tank is adequate to keep the concentration of the salt in the tank uniform, how much salt is in the tank at the end of 50 minutes? Assume that the density of the salt solution is essentially the same as that of water.

6.5 A storage tank contains 10 000 kg of a solution containing 5% acetic acid. A fresh feed of 500 kg per minute of pure water is entering the tank and dilutes the solution in the tank. The mixture is stirred well and the product leaves the tank at the rate of 500 kg per minute. At what instant of time will the acid concentration in the tank drop to 1%?

7 Energy Balance

Energy balance is the accounting of energy change accompanying a process. When dealing with chemical processes, we are more concerned with heat or thermal energy. Energy balance must show how much of energy is supplied, how much is used up in the different steps of the process, how much if any is obtained from some reaction in the process, and finally, what is left over. In most of the cases it will be a matter of determination of total energy requirement of the process in order to provide the correct amount of energy for the process.

$$\text{Energy input} = \text{Energy output} + \text{Energy losses}$$

For the accounting of energy, that is, heat, energy has to be quantified.

The unit of heat or enthalpy is the joule (J). The *heat capacity* is the heat required to raise the temperature of 1 kg of substance by 1 K. The unit of heat capacity is kJ kg^{-1} K^{-1}. The *specific heat capacity* of a substance is the ratio of the heat capacity of the substance to that of water. For solids and liquids, the heat capacity is expressed on mass basis; for gases it is expressed mostly on molar basis and it is known as *molal* or *molar heat capacity*. The unit of heat capacity, will then be kJ $kmol^{-1}$ K^{-1}.

Heat changes are of two types

- Sensible heat changes
- Latent or hidden heat changes

When a *sensible heat change* occurs, there is no phase transformation. The heat change (ΔQ) can be sensed by a variation in the temperature (dT) of the system.

$$\Delta Q = mCdT \text{ or } nCdT$$

where m is the mass of the substance, n is the number of moles of the substance and C is the heat capacity of the substance.

When a *latent heat change* occurs, there is a phase change and there is no temperature variation.

$$\Delta Q = m\lambda \text{ or } n\lambda$$

where λ is the latent heat of vapourisation of the substance. When mass is used in computations, C and λ should be expressed on mass basis, and when mole is used in computations, C and λ should be expressed on molar basis.

ENTHALPY CHANGES FOR SYSTEMS INVOLVING SENSIBLE HEAT CHANGES

Solid Systems

$\Delta Q = mC_s dT$, where C_s is the heat capacity of the solid.

Liquid Systems

$\Delta Q = mC_l dT$, where C_l is the heat capacity of the liquid.

For both systems, if C is temperature dependent,

$$C_s \text{ or } C_l = a + bT$$

where a and b are constants.

$$\begin{aligned}\Delta H &= m\int_{T_1}^{T_2} (a + bT)dT \\ &= m[a(T_2 - T_1) + b/2(T_2^2 - T_1^2)]\end{aligned}$$

where T_1 is the initial temperature and T_2 is the final temperature.

Gas systems

ΔQ depends on the route chosen, either at constant volume or constant pressure.

At *constant volume* the entire energy supplied goes to alter the internal energy as no work is done on the gas.

$$\Delta Q = \Delta E = nC_v dT$$

C_v is the heat capacity of the system under constant volume.

At *constant pressure* the energy supplied goes to increase the internal energy and also supplies energy required to do work on the gas.

$$\begin{aligned}\Delta Q = \Delta H &= \Delta E + pdv \\ &= nC_p dT\end{aligned}$$

C_p is the heat capacity of the system under constant pressure.

If C_p is temperature dependant, then

$$\begin{aligned}C_p &= a + bT + cT^2 \\ \Delta H &= n[a(T_2 - T_1) + b/2(T_2^2 - T_1^2) + c/3(T_2^3 - T_1^3)]\end{aligned}$$

Solid mixtures

$$\Delta Q = mC_{s(\text{mix})}dT$$

$$C_{s(\text{mix})} = \Sigma m_i C_{s(i)}$$

where m_i is the mass fraction and $C_{s(i)}$ is the heat capacity of the ith constituent.

Liquid mixtures

$$\Delta Q = mC_{l(\text{mix})}dT$$

$$C_{l(\text{mix})} = \Sigma m_i C_{l(i)},$$

where m_i is the mass fraction and $C_{l(i)}$ is the heat capacity of the ith constituent.

Gaseous mixtures

$$\Delta Q = nC_{p(\text{mix})}dT$$

$$C_{p(\text{mix})} = \Sigma n_i C_{p(i)},$$

where n_i is the number of moles and C_{pi} is the heat capacity of the of ith constituent.
C_p value at 101.325 kPa is denoted by C_p^o.

Mean heat capacity

If C_p is temperature dependent,

$$\begin{aligned}
C_p &= a + bT + cT^2 \\
\Delta H &= nC_p dT \\
&= n(a + bT + cT^2)dT \\
&= n[a(T_2 - T_1) + b/2(T_2^2 - T_1^2) + c/3(T_2^3 - T_1^3)]
\end{aligned}$$

If $C_{p(m)}$ is the *mean heat capacity* for the temperature range $T_1 - T_2$.

$$\begin{aligned}
\Delta H &= nC_{p(m)}(T_2 - T_1) \\
\Delta H &= nC_{p(m)}(T_2 - T_1) = n[a(T_2 - T_1) + b/2(T_2^2 - T_1^2) + c/3(T_2^3 - T_1^3)] \\
C_{p(m)} &= 1/(T_2 - T_1)\int_{T_1}^{T_2} C_p dT \\
&= [a(T_2 - T_1) + b/2(T_2^2 - T_1^2) + c/3(T_2^3 - T_1^3)]/(T_2 - T_1)
\end{aligned}$$

When the difference between T_1 and T_2 is not large, the mean heat capacity

$$C_{p(m)} = (C_p \text{ at } T_1 + C_p \text{ at } T_2)/2$$

where C_p at $T_1 = a + bT_1 + cT_1^2$, and C_p at $T_2 = a + bT_2 + cT_2^2$.

When the values of absolute enthalpies are known, then the enthalpy change may be computed by subtracting absolute enthalpy of the system at lower temperature from the absolute enthalpy of the system at higher temperature.

$$\Delta H = H \text{ at } T_2 - H \text{ at } T_1$$

where H at T_2 is the absolute enthalpy of the system at the higher temperature and H at T_1 is the absolute enthalpy of the system at the lower temperature.

ILLUSTRATIONS

7A The temperature of pure oxygen is raised from 350 to 1 500 K. Calculate the amount of heat to be supplied for raising the temperature of 1 kmol of oxygen using C_p data and absolute enthalpy data.

$$C_p = 26.03 + 11.75 \times 10^{-3}T - 2.34 \times 10^{-6}T^2 \text{ J mol}^{-1} \text{ K}^{-1}, \text{ where } T \text{ is in } K.$$

Absolute enthalpy of oxygen at 1 500 K, 400 K and 300 K are 49 273, 8 597 and 11 603 kJ kmol^{-1} respectively.

Solution

Amount of heat to be supplied,

$$\Delta H = nC_p dT$$

$$= n \int_{T_1}^{T_2} (a + bT + cT^2) dT$$

$$= n[a(T_2 - T_1) + b/2(T_2^2 - T_1^2) + c/3(T_2^3 - T_1^3)],$$

where T_1 and T_2 are the initial and final temperatures respectively in kelvin. Substituting the relevant values in the above equation,

$$\Delta H = 26.03(1\,500 - 350) + (11.75/2)(1\,500^2 - 350^2) - (2.34 \times 10^{-6}/2)(1\,500^3 - 350^3)$$

$$= 26.03(1150) + (117.5/2)(15^2 - 3.5)^2 - (2.34/3)(15^3 - 3.5^3) = 39\,834.5 \text{ kJ}$$

ΔH can be found by subtracting absolute enthalpy value of the system at the lower temperature from the absolute value of the system at the higher temperature. Absolute value of enthalpy at 350 K is the mean of the absolute enthalpy values at 300 K and 400 K.

$$\Delta H = \text{Absolute enthalpy at 1 500 K} - \text{Absolute enthalpy at 350 K}$$
$$= 49\,273 - 10\,100 = 39\,173 \text{ kJ}$$

7B Calculate the heat input to raise the temperature of 128 kg of sulphur dioxide from 373 K to 1 273 K. Perform the calculation in the following ways: (i) by integrating the expression, (ii) by using C_p mean value, and (iii) by using C_p value at mean temperature.

$$C_p \text{ of } SO_2 = 6.85 + 8.533 \times 10^{-3} T - 2.475 \times 10^{-6} T^2 \text{ cal mol}^{-1} \text{ K}^{-1}$$

Solution

(i) Heat to be supplied,

$$\Delta H = nC_p dT$$
$$= n \int_{T_1}^{T_2} (a + bT + cT^2) dT$$
$$= n[a(T_2 - T_1) + b/2(T_2^2 - T_1^2) + c/3(T_2^3 - T_1^3)]$$

where T_1 and T_2 are initial and final temperatures respectively in Kelvin.

Molecular weight of sulphur dioxide is 64. So 128 kg of sulphur dioxide = 2 kmol of sulphur dioxide.

Substituting the relevant values in the above equation,

$$\Delta H = 2\{(6.85 \times 900) - (8.533 \times 10^{-3}/2)(1\,273^2 - 373^2)$$
$$-(2.475 \times 10^{-6}/3)(1\,273^3 - 373^3)\}$$
$$= 2(11\,815) = 23\,630 \text{ kcal}$$

(ii) $$C_{p(m)} = 1/(T_2 - T_1) \int_{T_1}^{T_2} C_p dT$$

Substituting the relevant values in the above equation,

$$C_{p(m)} = \frac{11\,815}{900} = 13.13 \text{ kcal kmol}^{-1} \text{ K}^{-1}$$

$$\Delta H = nC_{p(m)} \Delta T = 2 \times 13.13 \times 900 = 23\,634 \text{ kcal}$$

(iii) $T_{mean} = [(1\,273 + 373)/2] = 823\,K$

$$C_p \text{ at } 823\,K = 6.85 + (8.533 \times 10^{-3} \times 823) - 2.475 \times 10^{-6} \times 823^2$$
$$= 12.194\,\text{kcal kmol}^{-1}\,K^{-1}$$

$$\Delta H = nC_{p\,(823)}\,\Delta T$$
$$= 2 \times 12.194 \times 900 = 21\,948\,\text{kcal}$$

7C Calculate the heat given off when 2 m^3 of nitrogen (measured at NTP) is cooled from 700°C to 20°C at constant pressure. C_p of nitrogen $= 6.386 + 1.762 \times 10^{-3}T - 0.265\,6 \times 10^{-6}T^2$, where T is in K and C_p is in cal mol^{-1} K^{-1}.

Solution

Number of moles, n = Volume ÷ molar volume (both at NTP) $= \dfrac{2}{22.4}$

$$\Delta H = \frac{2}{22.4}\int_{293}^{973} (6.386 + 1.76 \times 10^{-3}T - 0.2\,656 \times 10^{-3}T^2)\,dT$$

$$= 501.91\,\text{kcal}$$

7D How many joules of energy are needed to heat 24 kg of ozone from 273 K to 373 K? C_p of ozone is $34.33 + 42.86 \times 10^{-3}T - 13.21 \times 10^{-6}T^2$ J mol^{-1} K^{-1}, where T is in kelvin.

Solution

Molecular weight of ozone is 48.

Amount of ozone in 24 kg = 0.5 kmol

$$\Delta H = n\int_{273}^{373} C_p dT$$

$$= 0.5\,[34.33(373 - 273) + (428.6/2)(3.73^2 - 2.73^2)$$
$$-(13.21/3)(3.73^3 - 2.73^3)]$$
$$= 2\,892.5\,J$$

7E 66.26 m^3 of carbon dioxide gas at 2 atm pressure and 323 K has to be heated to 523 K. C_p of carbon dioxide = $6.34 + 10.14 \times 10^{-3}T - 3.42 \times 10^{-6}T^2$ cal mol^{-1} K^{-1} and T is given in K. Compute the heat duty in the following modes: (i) by integrating the expression for C_p, (ii) by using mean C_p value, and (iii) by using C_P at the mean temperature.

Solution

Amount of carbon dioxide is given by $n = \dfrac{PV}{RT} = \dfrac{2 \times 66.26}{0.082\,05 \times 323} = 5$ kmol

$$\text{(i)}\quad \Delta H = n\int_{T_1}^{T_2}(a + bT + cT^2)dT$$

$$= 5\,[6.34(200) + (101.4/2)(5.23^2 - 3.23)^2 - (3.42/3)(5.23^3 - 3.23^3)]$$

$$= 10\,005.85 \text{ kcal}$$

$$\text{(ii)}\quad C_{p(m)} = \frac{\Delta H}{n\Delta T} = \frac{10\,005.85}{5 \times 200} = 10.006 \text{ kcal kmol}^{-1}\text{ K}^{-1}$$

$$\Delta H = nC_{p(m)}\Delta T$$

$$= 5 \times 10.006 \times 200$$

$$= 10\,006 \text{ kcal}$$

(iii) C_p at the mean temperature, that is, 423 K

$$= 6.34 + 10.14 \times 10^{-3} \times 423 - 3.42 \times 10^{-6} \times 423$$

$$= 10.02 \text{ k cal kmol}^{-1}\text{ K}^{-1}$$

$$\Delta H = n \times C_p \text{ at mean temperature } \times \Delta T$$

$$= 5 \times 10.02 \times 200$$

$$= 10\,020 \text{ kcal}$$

7F A stream flowing at a rate of 500 mol per minute containing 35% nitrogen and 65% hydrogen is to be heated from 25°C to 200°C. Calculate the heat that must be transferred.
C_p of nitrogen = $5.4 + 1.2 \times 10^{-3}T$; C_p of hydrogen = $7.8 + 2.6 \times 10^{-3}T$, where C_p is in cal mol^{-1}K^{-1} and T is in K.

Solution

Basis: 500 mol of the mixture

The mixture contains 35mol% of nitrogen and the rest hydrogen.

Amount of nitrogen in the mixture $= \dfrac{500 \times 35}{100} = 175\,\text{mol}$

Amount of hydrogen in the mixture $= 500 - 175 = 325\,\text{mol}$

Heat required by nitrogen $= n \displaystyle\int_{T_1}^{T_2} (a + bT)dT$

$$= 175 \int_{298}^{473} (5.4 + 1.2 \times 10^{-3} T)dT$$

$$= 175[5.4(473 - 298) + (1.2 \times 10^{-3}/2)(473^2 - 298^2)]$$

$$= 176.66\,\text{kcal}$$

Similarly, heat required by hydrogen $= 325 \displaystyle\int_{298}^{473} (7.8 + 2.6 \times 10^{-3} T)dT$

$$= 325\,\{7.8(473 - 298) + (2.6 \times 10^{-3}/2)(473^2 - 298^2)\}$$

$$= 489.13\,\text{kcal}$$

Total heat transferred $= 176.66 + 489.13 = 665.79\,\text{kcal}$

7G Liquid benzene at 30°C is mixed and dissolved in liquid toluene at 100°C in the molar proportion 3:2. If there is no heat loss to the surroundings and the heat of mixing is zero, what will be the temperature of the solution after mixing is complete?

Table 7.1 Specific heat data: $C_p = a + bT$ kcal kg^{-1} ^{0}C^{-1}

Temperature (°C)	Specific heat of benzene	Specific heat of toluene
10	0.38	0.364
65	0.482	–
85	–	0.534

Solution

Based on the data given in Table 7.1, temperature dependent expressions for C_p in the form of $a + bT$ for benzene and toluene are obtained as given below. Here a is the base value at the lower temperature and b is the increase in C_p value for one degree increase in temperature.

Specific heat of benzene $= 0.38 + [(0.482 - 0.38)/55]T$

$$= 0.38 + 1.85 \times 10^{-3} T$$

Specific heat of toluene $= 0.364 + [(0.534 - 0.364)/75]T$
$= 0.364 + 2.27 \times 10^{-3}T$

As heat capacity values are known on a mass basis, molar ratio of benzene to toluene has to be changed to mass ratio. This is done by multiplying the number of moles of each by the respective molecular weight. Molecular weights of benzene and toluene are 78 and 92 respectively.

Benzene:Toluene $= 3:2$ by mole
$= 3 \times 78 : 2 \times 92 = 234 : 184 = 1.272 : 1$ by weight

Let the final temperature be T. Benzene, which is at a lower temperatue, gains heat and toluene loses heat. As there is no heat of mixing and no heat is lost to the surroundings, heat gained by benzene and that lost by toluene are numerically equal.

Heat lost by toluene $= 1 \int_{T}^{100} (0.364 + 2.27 \times 10^{-3}T)dT$

Heat gained by benzene $= 1.272 \int_{30}^{T} (0.38 + 1.85 \times 10^{-3}T)dT$

Equating and solving, $T = 75.05°\text{C}$

7H Flue gases leaving the stack of a boiler at 523 K have the following molar composition: 11.31% carbon dioxide, 13.04% water vapour, 2.17% oxygen, 73.48% nitrogen. Calculate the heat loss for 1 kmol of gas mixture above 298 K using heat capacity data. Also calculate the average heat capacity of dry flue gases.

C_p value for carbon dioxide $= 19.02 + 79.63 \times 10^{-3}T - 73.7 \times 10^{-6}T^2$ kJ kmol^{-1}K^{-1}
C_p value for water vapour $= 34.05 - 9.65 \times 10^{-3}T + 32.99 \times 10^{-6}T^2$ kJ kmol^{-1}K^{-1}
C_p value for oxygen $= 29.88 - 11.38 \times 10^{-3}T + 43.78 \times 10^{-6}T^2$ kJ kmol^{-1}K^{-1}
C_p value for nitrogen $= 29.41 - 3.00 \times 10^{-3}T + 5.45 \times 10^{-6}T^2$ kJ kmol^{-1}K^{-1}

Solution

For a gaseous mixture the heat changes can be calculated as given below:

$$\Delta H = nC_{P(\text{mix})}dT$$
$$C_{p(\text{mix})} = \Sigma n_i C_{p(i)}$$

where n_i is the mole fraction of the ith constituent and $C_{p(i)}$ is the heat capacity of the of ith constituent.

If C_p is given as $a + b \times 10^{-3}T + c \times 10^{-6}T^2$, then $C_{p(\text{mix})} = \sum n_i a_i + (\sum n_i b_i) \times 10^{-3}T + (\sum n_i c_i) \times 10^{-6}T^2$. C_p of the gaseous mixture is found by substituting the relevant values. Values given as mole per cent are converted to mole fraction by dividing by 100.

$$
\begin{aligned}
C_{p(\text{mix})} &= (0.1131 \times 19.02 + 0.1304 \times 34.05 + 0.0217 \times 29.88 + 0.7348 \times 29.41) \\
&\quad + (0.1131 \times 79.63 - 0.1304 \times 9.65 - 0.0217 \times 11.38 - 0.7348 \times 3.00) \times 10^{-3}T \\
&\quad + (-0.1131 \times 73.7 + 0.1304 \times 32.99 + 0.0217 \times 43.78 + 0.7348 \times 5.45) \times 10^{6}T^2 \\
&= 28.85 + 5.546 \times 10^{-3}T + 0.923 \times 10^{-6}T^2
\end{aligned}
$$

Then the heat change is calculated as in the previous problems.

$$
\begin{aligned}
\Delta H &= 28.85(523 - 298) + (55.46/2)(5.23^2 - 2.98^2) + (0.923/3)(5.23^3 - 2.98^3) \\
&= 7039.37\,\text{kJ}
\end{aligned}
$$

Composition of dry gas is worked out on a mole fraction basis leaving aside the water vapour content and is given in Table 7.2.

Table 7.2 Composition of dry gas

Constituent	**Moles (mol)**	**Mole fraction**
CO_2	11.31	0.13
O_2	2.17	0.025
N_2	73.48	0.845
	86.96	

$C_{p(\text{mix})}$ for dry flue gas is calculated as above and the heat change is then computed. This value on being divided by the temperature change gives the average C_p value of the dry flue gas.

$$
\begin{aligned}
C_{p\,(\text{mix})} &= (0.13 \times 19.02 + 0.025 \times 29.88 + 0.845 \times 29.41) \\
&\quad + (0.13 \times 79.63 - 0.025 \times 11.38 - 0.845 \times 3) \times 10^{-3}T \\
&\quad + (0.13 \times 73.7 + 43.78 \times 0.025 + 5.45 \times 0.845) \times 10^{-6}T^2 \\
&= 28.07 + 7.53 \times 10^{-3}T - 3.88 \times 10^{-6}T^2\ \text{kJ kmol}^{-1}\text{K}^{-1} \\
\Delta H &= 28.07(523 - 298) + (75.3/2)(5.23^2 - 2.98^2) \\
&\quad + (3.88/3)(5.23^3 - 2.98^3) \\
&= 6860.35\,\text{kJ}
\end{aligned}
$$

$$
\text{Average value of } C_{p\,(\text{mix})} = \frac{6890.35}{225} = 30.49\ \text{kJ kmol}^{-1}\text{K}^{-1}
$$

7I The gas from a sulphur burner has the following composition by volume: sulphur dioxide 7.8%, sulphur trioxide 0.8%, oxygen 12.2% and nitrogen 79.2%. Calculate the amount of heat that must be added to 1000 m^3 of this gas at STP to raise the temperature from 25°C to 1000°C.

Table 7.3 Mean molar heat capacity between 0°C to T°C in kcal $kmol^{-1}K^{-1}$

Gas	25°C	1000°C
SO_2	9.54	12.15
SO_3	12.11	17.86
O_2	7.02	7.94
N_2	6.96	7.57

Solution

Basis: 1000 m^3 of gas at STP (298 K and 1 atmosphere)

Number of kilomoles of gas, $n = \dfrac{PV}{RT} = \dfrac{1 \times 1000}{0.082\,05 \times 298} = 40.90\,\text{kmol}$

Amount of sulphur dioxide in the mixture $= 40.90 \times 0.078 = 3.19$ mol

Amount of sulphur trioxide in the mixture $= 40.90 \times 0.008 = 0.33$ mol

Amount of oxygen in the mixture $= 40.90 \times 0.122 = 4.99$ mol

Amount of nitrogen in the mixture $= 40.90 \times 0.792 = 32.39$ mol

Enthalpy of the gaseous mixture at 25°C relative to 0°C $= (\sum n_i C_{p\,(\text{mean})})\Delta T$, where n_i is the mole fraction of the ith constituent and $C_{p(mi)}$ is the mean heat capacity of the of ith constituent.

$$= [(3.19 \times 9.54) + (0.33 \times 12.11) + (4.99 \times 7.02) + (32.39 \times 6.96)](25 - 0) = 7\,372.325\,\text{kcal} \quad (1)$$

Enthalpy of gaseous mixture at 1000°C relative to 0°C $= (\sum n_i C_{p(m)})\Delta T$, where n_i is the mole fraction of the ith constituent and $C_{p(mi)}$ is the mean heat capacity of the of ith constituent.

$$= [(3.19 \times 12.15) + (0.33 \times 17.86) + (4.99 \times 7.94) + (32.39 \times 7.57)](1000 - 0) = 329\,465\,\text{kcal} \quad (2)$$

Amount of heat required to raise the temperature of gaseous mixture from 25°C to 1000°C $= (2) - (1) = 322\,092.675$ kcal

ENTHALPY CHANGES FOR SYSTEMS INVOLVING LATENT HEAT CHANGES AND PHASE TRANSFER

When a substance changes from one phase to another, heat is either absorbed or liberated without any change in temperature. The heat supplied for the change is termed as *latent heat* (λ), or *hidden heat*, as no change in temperature accompanies this.

Heat supplied to melt a solid to a liquid, or removed to convert a liquid back to a solid, is called the *latent heat of fusion* (λ_f).

Heat supplied to convert a liquid to vapour, or removed to condense a vapour to liquid, at constant pressure at the boiling point is called the *latent heat of vapourisation* (λ_v).

Heat supplied to convert a solid to vapour or removed to convert vapour to solid, is termed as the *latent heat of sublimation* (λ_s)

Values of λ are given in kJ kg^{-1} or kJ kmol^{-1}. When λ is given in kJ kmol^{-1}, it is known as the *molar latent heat*.

The heat change for a phase transfer is equal to $m\lambda$ kJ or $n\lambda$kJ, depending on the unit in which latent heat is expressed, where *m* refers to mass and *n* refers to the number of moles.

ILLUSTRATIONS

7J Calculate the heat added to 100 g of water to heat it from −20°C to 130°C at 1 atm pressure.

Specific heat of ice	$= 0.485$ cal g^{-1}°C^{-1}
Specific heat of water	$= 18.03$ cal g^{-1}°C^{-1}
Specific heat of steam	$= 7.256 + 2.298 \times 10^{-3} + 0.283 \times 10^6 T^2$ cal mol^{-1}°C^{-1}
Latent heat of fusion of ice	$= 1\,436$ cal mol^{-1}
Latent heat of vapourisation of water	$= 9717$ cal mol^{-1}

Solution

Basis: 100 g of water (100 g water = 100/18 mol)

Water is present as ice between −20°C and 0°C. Ice melts to water at 0°C. It remains as a liquid between 0°C and 100°C. Water boils at 100°C to become water vapour and remains in that state till it is heated to 130°C.

When there is a change in temperature, it is a sensible heat change (SH). When there is a phase change without temperature change it is a latent heat change (LH). SH requirement is given by $mC\Delta T$ and LH requirement is given by $m\lambda$ or $n\lambda$. Here *m* refers to mass, *n* refers to number of moles, *C* refers to heat capacity, λ refers to latent heat and ΔT refers to temperature difference.

Sensible heat required by ice $= 100 \times 0.485 \times 20 = 970$ cal

Latent heat required for conversion of ice to water $= (100/18) \times 1\,436 = 7\,977.8$ cal

Sensible heat required by water $= (100/18) \times 18.03 \times 100 = 10\,016.7$ cal

Latent heat required for conversion of water to water vapour $= (100/18) \times 9\,717 = 53\,983.3$ cal

Sensible heat required by water vapour

$$= (100/18) \int_{373}^{403} 7.256 + 2.298 \times 10^{-3} T + 0.283 \times 10^{-6} T^2) dT$$

$= 1\,370$ cal

Total heat required $= 74\,317.8$ cal

7K Calculate the heat required to raise the temperature of 100 kg of zinc from 0°C to 1200°C at 1 atm. Zinc melts at 419°C and boils at 907°C under atmospheric pressure. The mean heat capacity of zinc between 0°C and 419°C is 0.105 kcal kg^{-1} $°C^{-1}$, mean heat capacity of zinc between 419°C and 907°C is 0.109 kcal kg^{-1} $°C^{-1}$. λ_f is 1 660 kcal $kmol^{-1}$, λ_v is 26 900 kcal $kmol^{-1}$ at its normal boiling point and the mean heat capacity of zine vapour between 907°C and 1200°C is 4.97 kcal $kmol^{-1}$. Atomic weight of zinc = 65.4.

Solution

Basis: 100 kg of zinc (100 kg Zn = 100/56.4 = 1.529 kmole)

Zinc is present as a solid between 0°C and 419°C. It melts at 419°C. It remains a liquid between 419°C and 907°C. Liquid zinc boils at 907°C to become zinc vapour and remains in that state till it is heated to 1200°C.

When there is a change in temperature, it is a sensible heat change (SH). When there is a phase change without temperature change it is a latent heat change (LH). SH requirement is given by $mC\Delta T$ and LH requirement is given by $m\lambda$ or $n\lambda$. Here m refers to mass, n refers to mole, C refers to heat capacity, λ refers to latent heat, and ΔT refers to temperature difference.

Sensible heat required by zinc(s) $= 100 \times 0.105 \times 419$ $= 4\,399.5$ kcal

Latent heat required for conversion of zinc(s) to zinc(l) $= 1.529 \times 1\,660$ $= 2\,538.1$ kcal

Sensible heat required by zinc(l) $= 100 \times 0.109 \times 488$ $= 5\,319.2$ kcal

Latent heat required for conversion of zinc(l) to zinc(v) $= 1.529 \times 26\,900 = 41\,130.1$ kcal

Sensible heat required by zinc vapour $= 1.529 \times 4.97 \times 293$ $= 2\,226.5$ kcal

Total heat required $= 55\,613.4$ kcal

7L Based on a datum temperature of 0°C, calculate the enthalpy of humid air at 260°C and 1 atm pressure. The humid air contains 0.08 kg of water vapour per kilogram of dry air. Its dew point is 50°C.

Latent heat of vapourisation of water at 50°C = 570 kcal kg^{-1}
Heat capacity of water (v) = 0.45 kcal kg^{-1}°C^{-1}
Heat capacity of dry air = 0.25 kcal kg^{-1}°C^{-1}

Solution

Basis: 1 kg of dry air

Humid air contains 0.08 kg of water vapour. Upto the dew point water vapour is present in the liquid state. It vapourises at the dew point and is super heated above the dew point.

When there is a change in temperature, it is a sensible heat change (SH). When there is a phase change without temperature change it is a latent heat change (LH). SH requirement is given by $mC\Delta T$ and LH requirement is given by $m\lambda$. Here m refers to mass, C refers to heat capacity, λ refers to latent heat and ΔT refers to temperature difference.

Sensible heat required by water (l) $= 0.08 \times 1 \times (50 - 0) = 4$ kcal

Latent heat required for conversion of water (l) to water (v) $= 570 \times 0.08 = 45.60$ kcal

Sensible heat required by water (v) $= 0.08 \times 0.45 \times (260 - 50) = 7.56$ kcal

Sensible heat for dry air $= 1 \times 0.25 \times 260 = 62.40$ kcal

Enthalpy of the humid air at 260°C relative to 0°C is the sum of all the above = 119.56 kcal per kg of dry air

7M Saturated steam at 523 K is used to heat a batch reaction vessel. The reaction absorbs 550 kcal of heat per kilogram of material in the reactor. Heat loss from the steam chest to the surrounding is 1260 kcal per hour. The reactants are placed in the vessel at 294 K and at the end of the reaction the material is at 373 K. If the charge contains 150 kg of material and both feed and product have an average heat capacity of 0.77 kcal kg^{-1}K^{-1}, how much steam is needed per kilogram of charge? The charge remains in the reactor for one hour. Latent heat of vapourisation of steam is 410 kcal kg^{-1}.

Solution

Basis: One hour operation

150 kg of material is charged to the reactor at 294 K. The charge remains in the reactor for one hour and undergoes the reaction. The products leave the reactor at 373 K. The reaction requires heat of 550 kcal per kilogram of reactant.

Heat to be supplied = $150 \times 0.77 \times (373 - 294) = 9\,124.5$ kcal
Heat of reaction = 550×150 = 82 500 kcal
Heat lost = 1 260 kcal
Total heat required = 92 884.5 kcal

This is supplied by the condensation of steam, which releases 410 kcal kg^{-1}.

Steam requirement per hour = $92\,884.5 \div 410$ = 226.55 kg

ENTHALPY CHANGES FOR SYSTEMS INVOLVING CHEMICAL REACTIONS

When chemical reactions occur, heat is absorbed or evolved. Three types of heat changes are of interest.

1. Heat of combustion
2. Heat of formation
3. Heat of reaction

Heat of formation is defined as the heat absorbed or evolved when one mole of a compound is formed from its elements at a give temperature and pressure. A temperature of 25°C and pressure of 1 atm is referred to as the standard state.

Heat of combustion is the heat evolved when one mole of a substance undergoes complete combustion.

Heat of reaction is the heat evolved or absorbed when one or more substances react to form other compounds.

The heat of formation is a special type of heat of reaction from the elements. The reaction need not be a real one it can also be imaginary. It is normally evaluated experimentally, but can also be calculated indirectly from heat of combustion data.

It may be noted that the heat of formation of pure substances in their elemental state is zero.

ILLUSTRATIONS

7N When benzene is burnt completely to carbon dioxide(g) and water(l), the standard heat of combustion is 780 980 cal mol^{-1}. The standard heat of combustion of hydrogen(g) to water(g) is −68 317 cal mol^{-1} and that of carbon(s) to carbon dioxide (g) is −94 052 cal mol^{-1}. Calculate the standard heat of formation of benzene(l).

Solution

Combustion reactions for which data are available can be represented as follows:

$$C_6H_6(l) + 15/2\,O_2(g) \rightarrow 6CO_2(g) + 3H_2O(l) \qquad \Delta H_C^\circ = -780\,980 \text{ cal} \qquad (1)$$

$$1/2\,O_2(g) + H_2(g) \rightarrow H_2O(g) \qquad \Delta H_f^\circ = -68\,317 \text{ cal} \qquad (2)$$

$$C(s) + O_2(g) \rightarrow CO_2(g) \qquad \Delta H_f^\circ = -94\,052 \text{ cal} \qquad (3)$$

It is required to find the heat of formation of benzene, as per the reaction given below:

$$6C(s) + 3H_2(g) \rightarrow C_6H_6 \qquad (4)$$

This is obtained by the following scheme:

$$6(3) + 3(2) - (1) = (4)$$

Heat change for the required reaction:

$$6(-94\,052) + 3(-68\,317) - (-780\,980) = 11\,717 \text{ cal mol}^{-1}$$

70 Calculate the standard heat of formation of chloroform using the following heat of combustion data.

$$H_2(g) + 1/2\,O_2(g) \rightarrow H_2O(l) \qquad \Delta H_c^\circ = 68.32 \text{ kcal mol}^{-1} \qquad (1)$$

$$C(s) + O_2(g) \rightarrow CO_2(g) \qquad \Delta H_c^\circ = -98.05 \text{ kcal mol}^{-1} \qquad (2)$$

$$1/2\,H_2(g) + 1/2\,Cl_2(g) \rightarrow HCl(l) \qquad \Delta H_c^\circ = -40.02 \text{ kcal mol}^{-1} \qquad (3)$$

$$CHCl_3(g) + 1/2\,O_2(g) + H_2O(l) \rightarrow CO_2(g) + 3HCl(l) \qquad \Delta H_c^\circ = -121.8 \text{ kcal mol}^{-1} \qquad (4)$$

Solution

Heat of formation of $CH_3Cl(g)$ is given by

$$C(s) + 1/2\,H_2(g) + 3/2\,Cl_2(g) \rightarrow CHCl_3(g)$$

This is obtained by the following scheme:

$$(2) + 3(3) - (1) - (4)$$

$$\text{Heat change for the required reaction} = -94.05 + 3(-40.01) - (-68.32) - (-121.8)$$

$$= -23.99 \text{ kcal mol}^{-1}$$

Heat of reaction

It is the difference between the enthalpy of products and enthalpy of reactants. The standard heat of reaction can be calculated from the standard heat of formation.

Consider the reaction: $aA + bB \rightarrow cC + dD$.

The standard heat of reaction for this is given by

$$\Delta H_R^\circ = (\Delta H_f^\circ)_{\text{products}} - (\Delta H_f^\circ)_{\text{reactants}} = [c\Delta H_{f(C)}^\circ + d\Delta H_{f(D)}^\circ] - [a\Delta H_{f(A)}^\circ + b\Delta H_{f(B)}^\circ]$$

If ΔH_R° is positive, the reaction is endothermic, and if ΔH_R° is negative, the reaction is exothermic.

For reactions between organic compounds ΔH_R° can be calculated from ΔH_c° data.

$\Delta H_R^\circ = \Sigma \Delta H_c^\circ$ of reactants $- \Sigma \Delta H_c^\circ$ of products.

ILLUSTRATIONS

7P Calculate the standard heat of reaction for the formation of phosphorous as per the following equation.

$$Ca_3(PO_4)_2 + 3SiO_2 + 5C \rightarrow 3CaSiO_3 + 5CO + 2P$$

Heats of formation of $Ca_3(PO_4)_2$, SiO_2, $CaSiO_3$ and CO are −986.2, −205.4, −378.6 and −26.1 kcal mol^{-1} respectively. Find ΔH_f° of P by this reaction.

Solution

$\Delta H_R^\circ = \Delta H_f^\circ$ of products $- \Delta H_f^\circ$ of reactants

$= 1/2\,[3(-378.6) + 5(-26.1) + 2(0)] - [-986.2 + 3(205.4) + 5(0)]$

$= 1/2\,(-336.1)$ kcal per mol of P. It is also equal to ΔH_f° of P by this reaction.

7Q Calculate the standard heat of reaction of the following reaction.

$$2FeS(s) + 11/2\, O_2(g) \rightarrow Fe_2O_3(s) + 4SO_2(g)$$

The heats of formation of FeS_2, Fe_2O_3 and SO_2 are −42.52, −196.50 and −70.96 kcal mol^{-1}

Solution

The standard heat of reaction for this reaction is given by

$\Delta H_R^\circ = \Delta H_f^\circ$ of products $- \Delta H_f^\circ$ of reactants

$= [-196.5 + 4(-70.96)] - 2(-42.52)$

$= -395.3$ kcal mol^{-1}

7R Calculate the standard heat of reaction (ΔH_R°) of the following reaction:

$$C_2H_5OH(l) + CH_3COOH(l) \rightarrow C_2H_5O{\cdot}OC{\cdot}CH_3(l) + H_2O(l)$$

The standard heat of combustion of $C_2H_5OH(l)$, $CH_3COOH(l)$, $C_2H_5OOCCH_3(l)$ and $H_2O(l)$ are −326.7, −208.34, −538.76 and −68.32 kcal mol^{-1} respectively.

Solution

$$\Delta H_R^\circ = \Delta H_{c,\text{reactants}}^\circ - \Sigma \Delta H_{c,\text{products}}^\circ$$
$$= (-326.7 - 208.34) - (-538.76 - 68.32)$$
$$= 72.04\,\text{kcal mol}^{-1}$$

7S In burning limestone to produce lime, the reaction is

$$CaCO_3 \xrightarrow{\Delta} CaO + CO_2 \quad \Delta H = 19\,278\,\text{kcal kmol}^{-1}$$

If 25% of the heat of combustion of coal goes up the stack and 5% is lost by radiation, how many kilograms of coal of heating value 7 780 kcal kg^{-1} must be used to supply the heat of reaction of 1 tonne of limestone, assuming it to be 100% calcium carbonate.

Solution

Basis: One tonne of lime stone

Heat required for 100 kg calcium carbonate = 19 278 kcal

Heat required for 1000 kg of calcium carbonate = 192 780 kcal

25% of the heating value of coal is lost to the flue gas and 5% is lost by radiation.

Effective heat that can be recovered from 1 kg of coal $= 7\,780 \times \dfrac{70}{100} = 5\,446$ kcal

Coal required to supply the required heat $= 192\,780 \div 5\,446 = 35.4$ kg

Incomplete reactions

For incomplete reactions, the standard heat of reaction is calculated by subtracting the heats of formation of the reactants, which actually react, from the heats of formation of products that are actually formed.

ILLUSTRATIONS

7T In the production of metallic manganese, 12 kg of Mn_3O_4 were heated with 4 kg of coke. The resulting product was found to contain 5.2 kg of manganese, 2.8 kg MnO, unconverted charge and carbon monoxide gas. Calculate the standard heat of reaction of this process. ΔH_f° of Mn_3O_4, MnO, CO and coke are −331.4, −92.0, 2.64 and 2.6 kcal mol^{-1} respectively.

Solution

Basis: 12 kg of Mn_3O_4 and 4 kg of coke charged

The data given in the problem reveal that the reaction is an incomplete one. The standard heat of reaction for an incomplete reaction is calculated by subtracting the heats of formation of the reactants which actually react, from the heats of formation of products actually formed. So reactants actually reacted and products actually formed are to be found by a material balance.

The reactions taking place in the process are

$$Mn_3O_4 + 4C \rightarrow 3Mn + 4CO \qquad (1)$$

$$Mn_3O_4 + C \rightarrow 3MnO + CO \qquad (2)$$

Amount of Mn formed $= 5.2\,\text{kg} = 5.2 \div 55 = 0.0945\,\text{kmol}$

Amount of MnO formed $= 2.8\,\text{kg} = 2.8 \div 71 = 0.0394\,\text{kmol}$

Mn is formed in reaction (1) and MnO is formed in reaction (2). Consumption of 1 kmol of Mn_3O_4 leads to the formation of 3 kmol each of Mn and MnO.

Amount of Mn_3O_4 consumed $= [(0.0945 + 0.03940) \div 3] = 0.0446\,\text{kmol}$

For Mn formation 4 kmol of carbon, and for MnO formation, 1 kmol of carbon are consumed.

Amount of coke consumed $= \dfrac{0.0945 \times 4}{3} + (0.0394 \div 3) = 0.139\,\text{kmol.}$

1 kmol of CO is formed for the 1 kmol of carbon consumed.

Amount of CO formed $= 0.139\,\text{kmol}$

Heat of formation of reactants actually reacted

ΔH_f° of Mn_3O_4 $= 0.0446 \times (-331\,400) = -14\,780.4\,\text{kcal}$

ΔH_f° of coke $= 0.139 \times 2\,600 = 361.4\,\text{kcal}$

ΔH_f° of materials $= -14\,419\,\text{kcal}$

Heat of formation of products actually formed

ΔH_f° of MnO $= 0.0394 \times (-92\,000) = -3\,624.8$ kcal

ΔH_f° of Mn $= 0.0945 \times 0 = 0$ kcal

ΔH_f° of CO $= 0.139 \times (-26\,400) = -3\,669.6$ kcal

ΔH_f° of products $= -7\,294.4$ kcal

The standard heat of reaction for this reaction

$$\begin{aligned}\Delta H_R^\circ &= \Delta H_f^\circ \text{ of products actually formed} - \Delta H_f^\circ \text{ of reactants actually reacted}\\ &= -7\,294.4 - (-14\,419) \text{ kcal}\\ &= 7\,124.6 \text{ kcal}\end{aligned}$$

Effect of temperature on ΔH_R

If the reactants enter the reactor at T_1, while the products leave it at T_2, and if both T_1 and $T_2 > 298$ K, this can be achieved by

1. Cooling the reactants from T to 298 K—the enthalpy change for bringing the reactants to 298 K from $T_1 (-\Delta H_1)$
2. Carrying out the reaction at the state—standard heat of reaction at 298 K using standard heat of formation data (ΔH_R°)
3. Heating the products from 298 K to T_2K—the enthalpy change for taking the products from 298 K to $T_2 (+\Delta H_2)$

The algebraic sum of the three enthalpies gives the net energy change

$$-\Delta H_1 + \Delta H_R^\circ + \Delta H_2$$

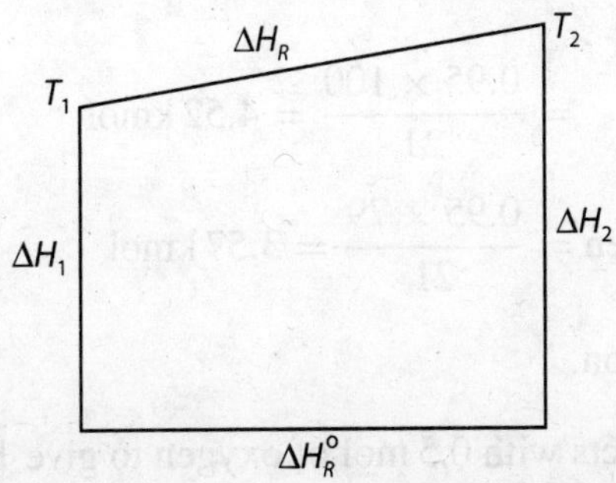

Fig. 7.1 Enyhalpy–temperature diagram

ILLUSTRATIONS

7U Carbon monoxide at 473 K is burnt with dry air at 773 K in 90% excess of that theoretically required. The products of combustion leave the reaction chamber at 1 273 K. Calculate the heat evolved in the reaction chamber in kilocalories per kilomole of carbon monoxide burnt, assuming complete combustion

$CO(g) + 1/2\,O_2(g) \rightarrow CO_2(g) \qquad \Delta H_R^\circ = -67.64\ \text{kcal mol}^{-1}$

$C_{p(\text{mean})}$ of CO between 25°C and 200°C $= 7.017\ \text{cal mol}^{-1}\text{K}^{-1}$

$C_{p(\text{mean})}$ of air between 25°C and 500°C $= 7.225\ \text{cal mol}^{-1}\text{K}^{-1}$

$C_{p(\text{mean})}$ of CO_2 between 25°C and 1000°C $= 11.92\ \text{cal mol}^{-1}\text{K}^{-1}$

$C_{p(\text{mean})}$ of O_2 between 25°C and 1000°C $= 7.941\ \text{cal mol}^{-1}\text{K}^{-1}$

$C_{p(\text{mean})}$ of N_2 between 25°C and 1000°C $= 7.507\ \text{cal mol}^{-1}\text{K}^{-1}$

Solution

Basis: 1 kmol of carbon monoxide, as given in the problem

The reaction taking place during combustion is

$$CO(g) + 1/2\,O_2(g) \rightarrow CO_2(g)$$

Theoretical oxygen required = 0.5 kmol

90% excess oxygen is supplied.

$$\text{Oxygen supplied} = \frac{0.5 \times 190}{100} = 0.95\ \text{kmol}$$

The source of oxygen is air. So oxygen is accompanied by nitrogen. Air contains 21mol% oxygen and 79mol% nitrogen.

$$\text{Air supplied} = \frac{0.95 \times 100}{21} = 4.52\ \text{kmol}$$

$$\text{Nitrogen accompanying oxygen} = \frac{0.95 \times 79}{21} = 3.57\ \text{kmol}$$

The reaction goes to completion.

1 mol of carbon monoxide reacts with 0.5 mol of oxygen to give 1 mol of carbon dioxide.

Carbon diodide formed = 1.0 kmol

Excess oxygen = oxygen fed – oxygen consumed = 0.95 – 0.5 = 0.45 kmol

The composition of the reactants and products is given in Table 7.4.

Table 7.4 Composition of the reactants and products

Reactants		Products	
CO	1.0 kmol	CO_2	1.0 kmol
Air	4.52 kmol	Excess O_2	0.45 kmol
		N_2	3.57 kmol

Enthalpy of reactants (ΔH_r) relative to standard state

CO $\quad 1.0 \times 7.017 \times (200 - 25) = 1\,227$ kcal

Air $\quad 4.52 \times 7.225 \times (500 - 25) = 15\,520$ kcal

$\Sigma\Delta H_r \quad 1\,227 + 15\,520 = 16\,747$ kcal

Enthalpy of products (ΔH_p) relative to standard state

$CO_2 \quad 1 \times (1000 - 25) \times 11.92 = 11\,630$ kcal

$O_2 \quad 0.45 \times 7.941 \times (1000 - 25) = 3\,482$ kcal

$N_2 \quad 3.57 \times 7.507 \times (100 - 25) \times = 26\,150$ kcal

$\Sigma\Delta H_p \quad 11\,630 + 3\,482 + 26\,150 = 41\,262$ kcal

$$\text{Net heat change } (\Delta H) = \Sigma\Delta H_p + \Delta H_R^\circ - \Sigma\Delta H_r$$

$$= 41\,262 - 67\,640 - 16\,747$$

$$= -43\,125 \text{ kcal}$$

7V Sulphur dioxide gas is oxidised in 100% excess air with 80% conversion to sulphur trioxide. Gases enter the converter at 673 K and leave at 773 K. How much heat is absorbed in the heat exchanger of the converter per kilomole of sulphur dioxide fed? Standard heat of reaction (ΔH_R°) = −98.9 kJ mol^{-1} of SO_3. Mean C_p values of sulphur trioxide, sulphur dioxide, oxygen and nitrogen are 15,52, 11.08, 7.46 and 7.13 cal mol^{-1}K^{-1} respectively.

Solution

Basis: 1 kmol of sulphur dioxide fed

The reaction taking place during combustion is

$$SO_2 + 1/2\, O_2 \rightarrow SO_3$$

Theoretical oxygen required = 0.5 kmol

100% excess of oxygen is supplied.

$$\text{Oxygen supplied} = \frac{0.5 \times 200}{100} = 1.00 \text{ kmol}$$

The source of oxygen is air. So oxygen is accompanied by nitrogen. Air contains 21mol% oxygen and 79mol% nitrogen.

Amount of nitrogen accompanying oxygen $= \dfrac{1.00 \times 79}{21} = 3.76\,\text{kmol}$

The reaction goes to 80% completion. So only 0.8 kmol sulphur dioxide reacts.

1 mol of sulphur dioxide reacts with 0.5 mol oxygen to give 1 mol of sulphur trioxide.

Sulphur trioxide formed = 0.8 kmol

Oxygen consumed = 0.4 kmol

Excess oxygen = oxygen fed − oxygen consumed = 1.0 − 0.40 = 0.60 kmol

The composition of the input and output are given in Table 7.5.

Table 7.5 Composition of the input and output

Input	Kilomoles (kmol)	Output	Kilomoles (kmol)
SO_2	1.00	SO_2	0.20
O_2	1.00	Excess O_2	0.60
N_2	3.76	N_2	3.76
		SO_3	0.80

ΔH_R° is given in kilojoules and the mean C_p values in calories. The problem is solved on joule basis throughout. 1 cal = 4.187 J.

Enthalpy of the reactants relative to standard state

$$= [(1 \times 11.08) + (1 \times 7.45) + (3.76 \times 7.13)] \times (673 - 298)] \times 4.187$$
$$= -80\,564.3\,\text{kJ (the negative sign indicates that heat is released)}$$

$\Delta H_R^\circ = -98\,900 \times 0.8 = -79\,120\,\text{kJ}$ (because 0.8 kmol of sulphur dioxide react)

Enthalpy of the products relative to standard state

$$= [(0.8 \times 15.52) + (11.08 \times 0.2) + (7.46 \times 0.6) + (3.76 \times 7.13)] \times (773 - 298)]\,4.187$$
$$= +100\,793.4\,\text{kJ (the positive sign indicates that heat is absorbed)}$$

Net heat change

$$= -80\,564.3 - 79\,120 + 100\,793.4\,\text{kJ} = -58\,890.9\,\text{kJ}.$$

58 890.9 kJ of heat is absorbed by the heat exchanger.

7W Calculate the heat of reaction for the following reaction

$CO_2 + 4H_2 \rightarrow 2H_2O + CH_4$ at 500°C with the data given below.

Standard heat of reaction is $-39\,433$ kcal kmol^{-1}.

Table 7.6 Enthalpy values in kcal kmol^{-1}

Temperature°C	CO_2	H_2	H_2O	CH_4
25	218	172	200	210
500	5 340	3 499	4 254	5 730

Solution

The enthalpy change for heating a material from 25°C to 500°C is also the difference in absolute enthalpy at 500°C and 25°C.

Heat released by the reactants in cooling them from 500°C to 25°C ($\sum \Delta H_r$)

$$= (5\,340 - 218) + 4(3\,499 - 172) = -18\,430 \text{ kcal}$$

Standard heat of reaction $= -39\,433$ kcal

Heat absorbed in heating the product from 25°C to 500°C ($\sum \Delta H_p$)

$$= 2(4\,254 - 200) + (5\,730 - 210) = 13\,628 \text{ kcal}$$

Heat of reaction at 500°C is the algebraic sum of these three values

$$= -18\,430 - 39\,433 + 13\,628 = -44\,235 \text{ kcal}$$

7X The gases entering the converter of a sulphuric acid plant contain 2% sulphur trioxide, 8.5% sulphur dioxide, 9.6% oxygen and 79.8% nitrogen. No new material is added to the converter or removed from it. The outgoing gases contain 82.46% nitrogen. Calculate the heat removed per 100 kmol of entering gas if the reaction takes place isothermally at 723 K.

Standard heat of formation of sulphur dioxide and sulphur trioxide are $-70\,940$ and $-94\,390$ cal mol^{-1} respectively. Mean heat capacity values for sulphur dioxide, oxygen, sulphur trioxide and nitrogen between 298 K and 883 K are 12.32, 7.40, 10.91 and 7.00 cal mol^{-1}K^{-1} respectively.

Solution

Basis: 100 kmol of gases entering the converter

The reaction in the converter is

$$SO_2 + 1/2\, O_2 \rightarrow SO_3$$

1 mol of sulphur dioxide reacts with 0.5 mol of oxygen to give 1 mol of sulphur trioxide.

No material is added to or removed from the converter. So the reduction in the volume of outgoing gases is equal to the oxygen consumed and twice that will give the sulphur dioxide consumed, or sulphur trioxide formed.

The outgoing gases contain 82.46mol% nitrogen, which is equal to 79.8 kmol.

Amount of outgoing gases $= \dfrac{79.8 \times 100}{82.46} = 96.77\,\text{kmol}$

Oxygen consumed $= 100 - 96.77 = 3.23\,\text{kmol}$

Sulphur dioxide consumed or sulphur trioxide formed $= 3.23 \times 2 = 6.46\,\text{kmol}$

Composition of outgoing gases:

Amount of sulphur dioxide $= 8.5 - 6.46 = 2.04\,\text{kmol}$

Amount of oxygen $= 9.6 - 3.23 = 6.37\,\text{kmol}$

Amount of sulphur trioxide $= 2.1 + 6.46 = 8.56\,\text{kmol}$

Amount of nitrogen $= 79.8\,\text{kmol}$

If the reactants are assumed to enter at 298 K, the enthalpy of reactants are relative to the standard state $\Sigma \Delta H_r = 0$

Standard heat of reaction, $\Delta H_R^\circ = \Delta H_{f(SO_3)}^\circ - \Delta H_{f(SO_2)}^\circ$

$$= -94\,390 - (-70\,940) = -23\,450\,\text{kcal kmol}^{-1}\text{of } SO_3$$

$$= 6.46 \times (-23\,450) = -151\,487\,\text{kcal}$$

Enthalpy of products relative to standard state

$$= [(2.04 \times 12.32) + (6.37 \times 7.4) + (8.56 \times 10.91) + (79.8 \times 7.0)] \times (723 - 298)$$

$$= +307\,810\,\text{kcal}$$

Net heat change $= -321\,840 - 151\,487 + 307\,810$

$= -165\,517$ kcal. This much heat has to be removed

So 165 517 kcal of heat have to be removed from the converter.

Temperature of reaction

If a reaction proceeds without loss or gain of heat and if all the products of the reaction remain together in a single stream, the products will assume a definite temperature known as adiabatic reaction temperature (ART). The products here include all the materials in the final stream, that is, inerts, excess reactants and new compounds formed. If the reaction is incomplete, only the standard heat of reaction resulting from the degree of completion actually obtained is to be considered, and the products include some of each of the original reactants.

If the reactants enter the reactor at T_1 and T is the ART, then the heat change in the reaction chamber can be given as

$$\Delta H = \Sigma \Delta H_p + \Delta H_R^\circ - \Sigma \Delta H_r$$

$\Sigma \Delta H_p$ = Enthalpy of products relative to standard state

ΔH_R° = Standard heat of reaction

$\Sigma \Delta H_r$ = Enthalpy of reactants relative to standard state

Here $\Delta H = 0$ as the reaction is adiabatic.

$$\Sigma \Delta H_p = \Sigma \Delta H_r - (\pm \Delta H_R^\circ)$$

$$= \Sigma n_i \int_{298}^{T} C_{pi} dT = \Sigma n_i C_{p(\text{mean})} (T - 298)$$

$$= \Sigma n_i [a_i (T - 298) + (b_i/2)(T^2 - 298^2) + (c_i/3)(T^3 - 298^3)]$$

$$= \Sigma \Delta H_r - \Delta H_R^\circ$$

The adiabatic reaction temperature (ART), T, is obtained by solving the reaction. ART for a combustion reaction is known as the theoretical flame temperature (TFT). The temperature attained when a fuel is burnt in air or oxygen without gain or loss of heat is termed as TFT.

ILLUSTRATIONS

7Y Calculate the theoretical flame temperature for carbon monoxide burnt at constant pressure with 100% excess air. The reactants enter at 95°C. Standard heat of reaction (ΔH_R°) = −67 636 cal mol^{-1}. $C_{p(\text{mean})}$ values for air, carbon monoxide, carbon dioxide, oxygen and nitrogen are 6.993, 6.98, 12.94, 8.35 and 7.92 cal mol^{-1} ^{0}C^{-1} respectively.

Solution

Basis: 1 mol of carbon monoxide burnt

The reaction taking place during combustion is

$$CO(g) + 1/2\ O_2(g) \rightarrow CO_2(g)$$

Theoretical oxygen required = 0.5 kmol

100% excess oxygen is supplied

$$\text{Oxygen supplied} = \frac{0.5 \times 200}{100} = 1.00\ \text{kmol}$$

The source of oxygen is air. So oxygen is accompanied by nitrogen. Air contains 21mol% oxygen and 79mol% nitrogen.

$$\text{Air supplied} = \frac{1.0 \times 100}{21} = 4.76\ \text{kmol}$$

$$\text{Nitrogen accompanying oxygen} = \frac{1.00 \times 79}{21} = 3.76\ \text{kmol}$$

The reaction goes to completion.

1 mol of carbon monoxide reacts with 0.5 mol of oxygen to give 1 mol of carbon dioxide.

Carbon monoxide formed = 1.0 kmol

Oxygen consumed = 0.5 kmol

Excess oxygen = oxygen fed – oxygen consumed

= 1.00 – 0.5 = 0.50 kmol

The composition of the reactants and products is given in Table 7.7.

Table 7.7 Composition of the reactants and products

Reactants	Kilomoles (kmol)	Products	Kilomoles (kmol)
CO	1.0	CO_2	1.0
Air	4.76	Excess O_2	0.50
		N_2	3.76

Heat released when the reactants are cooled from 95°C to 25°C = [(1 × 6.981) + (6.993 × 4.76)](95 – 25)

= 2 819.72 cal

ΔH_R° = –67 636 cal (negative sign indicates that energy is released)

Total heat released = 70 455.72 cal

This should be equal to the enthalpy of the products relative to standard state. Let the TFT be T°C

$$70\,455.72 = [(1 \times 12.94) + (0.5 \times 8.35) + (3.76 \times 7.92)]\,(T - 25)$$

$$= 46.91\,(T - 25)$$

$$\therefore \quad (T - 25) = 1\,501.93\text{°C}$$

$$T = 1\,526.93\text{°C}$$

7Z Find the theoretical flame temperature of a gas containing 30% carbon monoxide and 70% nitrogen when burnt with 100% excess air. Both the reactants enter at 298 K. Standard heat of formation of carbon dioxide and carbon monoxide are $-393\,700\ \text{J mol}^{-1}$ and $-110\,600\ \text{J mol}^{-1}$ respectively.

Table 7.8 Mean C_p values in $\text{J mol}^{-1}\ \text{K}^{-1}$

Temperature (K)	CO_2	O_2	N_2
1 200	49.4	33.0	31.2
1 400	50.8	33.6	31.8

Solution

Basis: 1 mol of carbon monoxide fed to the burner

Compostion of the feed gas is 30% carbon monoxide and 70% nitrogen.

$$\text{Amount of nitrogen in the feed gas} = \frac{1 \times 70}{30} = 2.33\ \text{mol}$$

The reaction taking place is

$$CO + 1/2\ O_2 \rightarrow CO_2$$

Oxygen required $= 0.5$ mol

100% excess oxygen is supplied.

$$\text{Oxygen supplied} = \frac{0.5 \times 200}{100} = 1.00\ \text{mol}$$

The source of oxygen is air. So oxygen is accompanied by nitrogen. Air contains 21mol% oxygen and 79mol% nitrogen.

$$\text{Amount of nitrogen from air} = \frac{1.0 \times 79}{21} = 3.76\ \text{mol}$$

Amount of nitrogen in feed $= 2.33$ mol

Total nitrogen entering the burner $= 6.09$ mol

The reaction goes to completion.

1 mol of carbon monoxide reacts with 0.5 mol oxygen to give 1 mol of carbon dioxide.

Carbon monoxide formed $= 1.0$ kmol

Excess oxygen $=$ oxygen fed $-$ oxygen consumed

$= 1.00 - 0.5 = 0.50$ kmol

The composition of the reactants and products is given in Table 7.9.

Table 7.9 Composition of the reactants and products

Reactants	Kilomoles (kmol)	Products	Kilomoles (kmol)
CO	1.00	CO_2	1.00
O_2	1.00	O_2	0.50
N_2	6.09	N_2	6.09

As reactants enter at standard state,

$\sum \Delta H_r = 0$

ΔH_f° of $CO_2 = -393\,700$ J mol^{-1}

ΔH_f° of CO $= -110\,600$ J mol^{-1}

ΔH_f° of $O_2 = 0$ (as it is in the elemental state)

$\Delta H_R^\circ = \Delta H_f^\circ$ of $CO_2 - \Delta H_f^\circ$ of CO

$= -283\,100$ J

Let the theoretical flame temperature (TFT) be 1 400 K.

Heat absorbed in heating the products from standard state to TFT, $\Sigma \Delta H_p$

$= [(1 \times 50.8) + (0.5 \times 33.6) + (6.09 \times 31.8)](1\,400 - 298)$

$= 287\,909$ J

Let the TFT be 1 200 K.

Heat absorbed in heating the products from standard state to TFT, $\Sigma \Delta H_p$

$= [(1 \times 49.4) + (0.5 \times 33.0) + (6.09 \times 31.21)](1\,200 - 298)$

$= 230\,822$ J

At TFT, $\Sigma \Delta H_p$ must match numerically with ΔH_R°. So TFT must lie between 1 200 and 1400 K.

TFT $= 1\,200 + \{200[(283\,100 - 230\,822) \div (287\,909 - 230\,822)]\}$

$= 1\,383.15$ K

7AA Dry methane at 50°C is burnt with 50% excess dry air at 500°C. Assuming complete combustion, calculate the theoretical flame temperature. Standard heat of combustion of methane is -786.12 kJ mol^{-1}. Mean heat capacities of methane, air, carbon dioxide, oxygen, nitrogen and water vapour are 34.19, 28.94, 51.79, 32.40, 32.15 and 40.19 J mol^{-1} K^{-1} respectively.

Solution

Basis: 1 mol of methane fed to the burner

The reaction is

$$CH_4 + 2O_2 \rightarrow CO_2 + 2H_2O$$

Theoretical oxygen required $= 2.0$ mol

50% excess oxygen is supplied.

$$\text{Oxygen supplied} = \frac{2.0 \times 150}{100} = 3.0 \text{ mol}$$

The source of oxygen is air. So oxygen is accompanied by nitrogen. Air contains 21mol% oxygen and 79mol% nitrogen.

$$\text{Amount of nitrogen from air} = \frac{3.0 \times 79}{21} = 11.28 \text{ mol}$$

1 mol of methane consumes 2 mol of oxygen to give 1 mol of carbon dioxide and 2 mol of water.

Amount of carbon dioxide formed $= 1.0$ mol

Amount of water formed $= 2.0$ mol

Excess oxygen $=$ oxygen fed $-$ oxygen consumed

$= 3.0 - 2.0 = 1.0$ mol

The composition of the reactants and products is given in Table 7.10.

Table 7.10 Composition of the reactants and products

Reactants	Moles (mol)	Products	Moles (mol)
CH_4	1.0	CO_2	1
O_2	3.0	O_2	1
N_2	11.28	N_2	11.28
Air	14.28	Water vapour	2

Methane enters at 50°C and air enters at 500°C.

Heat released when the reactants are cooled to standard state ($\sum \Delta H_r$)

$$= [1 \times 34.19 \times (50 - 25)] + [14.28 \times 28.94 \times (500 - 25)]$$

$$= -197\,155\ \text{J}$$

Standard heat of reaction (ΔH_R°) $= -786\,120$ J

Assuming TFT to be T°C,

Heat absorbed when the products are heated to $T(\sum \Delta H_p)$

$$= [(1 \times 51.79) + (1 \times 32.4) + (11.28 \times 32.15) + (2 \times 40.19)]\ (T - 25)$$

$$= 526.22\ (T - 25)$$

As the reaction is adiabatic, the algebraic sum of all three should be equal to zero,

$$-197\,155 - 786\,120 + 526.22(T - 25) = 0$$

$$\therefore\ T = 1\,843.56°\text{C}$$

7BB Carbon monoxide is burnt with 100% excess air. Determine the final temperature obtained by the gaseous products of combustion, if combustion is adiabatic and complete. 20% of the heat is lost to the surroundings. Mean heat capacity of oxygen, nitrogen and carbon dioxide is 31.9, 32.15 and 51.8 J mol^{-1}K^{-1} respectively. Standard heat of reaction = −283.2 kJ kmol^{-1} for carbon monoxide.

Solution

Basis: 1 kmol of carbon monoxide burnt

The reaction taking place during combustion is

$$CO(g) + 1/2\ O_2(g) \rightarrow CO_2(g)$$

Theoretical oxygen required = 0.5 kmol

100% excess oxygen is supplied

$$\text{Oxygen supplied} = \frac{0.5 \times 200}{100} = 1.00\ \text{kmol}$$

The source of oxygen is air. So oxygen is accompanied by nitrogen. Air contains 21mol% oxygen and 79mol% nitrogen.

Amount of air supplied $= \dfrac{1.0 \times 100}{21} = 4.76\,\text{kmol}$

Amount of nitrogen accompanying oxygen $= \dfrac{1.00 \times 79}{21} = 3.76\,\text{kmol}$

The reaction goes to completion.

1 mol of carbon monoxide reacts with 0.5 mol oxygen to give 1 mol of carbon dioxide.

Amount of carbon dioxide formed $= 1.0\,\text{kmol}$

Excess oxygen = oxygen fed – oxygen consumed

$= 1.00 - 0.5 = 0.5\,\text{kmol}$

The composition of the reactants and products is given in Table 7.11.

Table 7.11 Composition of the reactants and products

Reactants	Kilomoles (kmol)	Products	Kilomoles (kmol)
CO	1.0	CO_2	1.0
Air	4.76	Excess O_2	0.50
		N_2	3.76

Assuming the reactants enter at standard state,

Heat released in cooling the reactants to standard state $(\sum \Delta H_r) = 0$

$\Delta H_R^\circ = -283\,200\,\text{J}$

20% of the heat is lost to the surroundings. Only, the balance heat can be realized.

Effective $\Delta H_R^\circ = -283\,200\,\text{J} \times \dfrac{80}{100} = -226\,560\,\text{J}$

Let the final temperature reacted by the products be T°C

Heat to be supplied to heat the products from standard state to T°C

$= [(1 \times 51.8) + (0.5 \times 31.9) + (3.76 \times 32.15)](T - 25)$

$= 188.63(T - 25)$

As the reaction is adiabatic this should be equal to the effective ΔH_R°

$188.63(T - 25) = 226\,560$

$\therefore \quad T = 1\,176.08°\text{C}$

7CC Determine the theoretical flame temperature of a gas containing 40% carbon monoxide, 35% hydrogen, 5% carbon dioxide and the rest nirogen, when burnt with 100% excess air. Both air and gas are initially at 25°C.
Mean molar heat capacity of the gas between 25°C and theoretical flame temperature for carbon dioxide, water vapour, oxygen and nitrogen are 12.7, 9.9, 8.2 and 7.8 cal $mol^{-1}K^{-1}$ respectively. ΔH_C° for carbon monoxide and hydrogen is −67 636 cal mol^{-1} and −68 317 cal mol^{-1} respectively.

Solution

Basis: 1 mol of the fuel gas

Composition of fuel gas

Amount of carbon monoxide in fuel = 0.40 mol

Amount of hydrogen in fuel = 0.35 mol

Amount of carbon dioxide in fuel = 0.05 mol

Amount of nitrogen in fuel = 0.20 mol

The combustion reactions are

$$CO + 1/2\, O_2 \rightarrow CO_2$$
$$H_2 + 1/2\, O_2 \rightarrow H_2O$$

1 mol of carbon monoxide consumes 0.5 mol of oxygen to form 1 mol of carbon dioxide.

1 mol of hydrogen consumes 0.5 mol of oxygen to form 1 mol of water.

Theoretical oxygen required $= (0.40 \div 2) + (0.35 \div 2) = 0.375$ mol

100% excess oxygen is supplied.

Oxygen supplied $= \dfrac{0.375 \times 200}{100} = 0.75$ mol

Source of oxygen is air. So oxygen is accompanied by nitrogen. Air contains 21mol% oxygen and 79mol% nitrogen.

Amount of nitrogen accompanying oxygen $= \dfrac{0.75 \times 79}{21} = 2.82$ mol

Total nitrogen entering the burner = nitrogen in air + nitrogen in fuel

$= 2.82 + 0.20 = 3.02$ mol

Carbon dioxide formed = 0.40 mol

Water formed = 0.35 mol

Excess oxygen = oxygen fed – oxygen consumed

$$= 0.75 - 0.375 = 0.375 \text{ mol}$$

The composition of the reactants and products is given in Table 7.12.

Table 7.12 Composition of the reactants and products

Reactants	Moles (mol)	Products	Moles (mol)
O_2	0.75	O_2	0.375
CO_2	0.4	CO_2	0.40
H_2	0.35	Water vapour	0.35
CO_2	0.05	N_2	3.02
N_2	3.02		

As reactants enter at 298 K, enthalpy of reactants relative to standard state, $\Sigma\Delta H_r = 0$

0.4 mol of carbon monoxide and 0.35 mol of hydrogen undergo combustion.

Standard heat of reaction, ΔH_R°

$$= 0.4 \times (-67\,636) + 0.35 \times (-68\,317) = -50\,965 \text{ cal}$$

Let the TFT be T K.

Enthalpy of products relative to standard state, $\Sigma\Delta H_P$

$$= [(0.45 \times 12.7) + (0.35 \times 9.9) + (0.375 \times 8.2) + (3.02 \times 7.8)](T - 298)] = 39.28(T - 298)$$

For this case, $\Delta H_R^\circ = \Sigma\Delta H_P$

$$39.28(T - 298) = 50\,965$$

$$\therefore \quad T = 1\,595.48 \text{ K}$$

7DD A gas containing 20% carbon monoxide and 80% nitrogen is burnt with 100% excess air, both air and gas initially being at 25°C. A flame temperature of 942°C is attained during combustion. Calculate the enthalpy of the products. Mean molar heat capacities between 25°C and 942°C for carbon dioxide, oxygen and nitrogen are 11.74, 7.9 and 7.45 cal mol^{-1}K^{-1} respectively.

Solution

Basis: 1 mol of carbon monoxide fed.

Feed contains 20% carbon monoxide and 80% nitrogen.

Amount of nitrogen accompanying carbon monoxide $= \dfrac{1 \times 80}{20} = 4$ mol

The reaction taking place during combustion is

$$CO(g) + 1/2\ O_2(g) \rightarrow CO_2(g)$$

Theoretical oxygen required $= 0.5$ mol.

100% excess oxygen is supplied

Amount of oxygen supplied $= \dfrac{0.5 \times 200}{100} = 1.00$ mol

The source of oxygen is air. So oxygen is accompanied by nitrogen. Air contains 21mol% oxygen and 79mol% nitrogen.

Amount of nitrogen accompanying oxygen $= \dfrac{1.00 \times 79}{21} = 3.76$ mol

Amount of nitrogen in feed $= 4.00$ mol

Total nitrogen entering the burner $= 7.76$ mol

The reaction goes to completion.

1 mol of carbon monoxide reacts with 0.5 mol of oxygen to give 1 mol of carbon dioxide.

Carbon dioxide formed $= 1.0$ mol

Oxygen consumed $= 0.5$ mol

Excess oxygen = oxygen fed – oxygen consumed

$= 1.00 - 0.50$

$= 0.50$ mol

The composition of the input and output are given in Table 7.13.

Table 7.13 Composition of input and output

Input	Moles (mol)	Output	Moles (mol)
CO	1.0	CO_2	1.0
O_2	1.0	O_2	0.5
N_2	7.76	N_2	7.76

Enthalpy of the products $= [(1 \times 11.74) + (0.5 \times 7.90) + (7.76 \times 7.45)](942 - 25)$

$= 66\,482.5$ cal per mole CO burnt $\equiv 13\,296.5$ cal per mole fuel fed

7EE Dry methane is burnt with dry air, both being initially at 25°C. The theoretical flame temperature is 1 570 K. If combustion is complete, find the excess air used. $\Delta H_R^\circ = -801\ \text{kJ mol}^{-1}$. $C_{p(\text{mean})}$ for carbon dioxides, water vapour, nitrogen and oxygen is 51.79, 40.19, 32.15 and 32.40 J mol^{-1}K^{-1} respectively.

Solution

Basis: 1 mol of methane burnt

The reaction is

$$CH_4 + 2O_2 \rightarrow CO_2 + 2H_2O$$

Theoretical oxygen required = 2.0 mol.

Let X be the number of moles of excess oxygen fed.

Oxygen supplied = $(2.0 + X)$ mol

The source of oxygen is air. So oxygen is accompanied by nitrogen. Air contains 21mol% oxygen and 79mol% nitrogen.

$$\text{Amount of nitrogen from air} = \frac{(2.0 + X) \times 79}{21} = 3.76(2 + X)\ \text{mol}$$

1 mol of methane consumes 2 mol of oxygen to give 1 mol of carbon dioxide and 2 mol of water.

Carbon dioxide formed = 1.0 mol

Water formed = 2.0 mol

Excess oxygen = oxygen fed – oxygen consumed

= $(2.0 + X) - 2.0 = X$ mol

The composition of the input and output are given in Table 7.14.

Table 7.14 Composition of input and output

Input	Moles (mol)	Output	Moles (mol)
CH_4	1.0	CO_2	1.0
O_2	$2.0 + X$	O_2	X
N_2	$3.76(2 + X)$	N_2	$3.76(2 + X)$
		Water vapour	2.0

As the reactants enter at 25°C, the enthalpy of the reactants relative to the standard state is zero. So standard heat of reaction is equal to enthalpy of products relative to standard state.

$$\Delta H_R^\circ = \Sigma \Delta H_p$$

$$801\,000 = [(1 \times 51.79) + (2 \times 40.19) + 32.4X + 32.15(7.52 + 3.76X)](1\,570 - 298)$$

$$\therefore \quad X = 1.646\,6 \text{ mol}$$

$$\text{Per cent excess of oxygen supplied} = \frac{\text{Excess} \times 100}{\text{Theoretical}} = \frac{1.646\,6 \times 100}{2} = 82.33\%$$

Problems for Practice

7.1 Calculate the amount of heat given off when 1 m^3 of air at NTP cools from 500°C to −100°C. C_p of air $= 6.39 + 1.76 \times 10^{-3}T - 0.27 \times 10^{-6}T^2$ cal mol^{-1} K^{-1}.

7.2 Calculate the heat input to raise the temperature of 132 kg of carbon dioxide from 100°C to 1000°C. Perform the calculation (i) by integrating the expression, (ii) by using the C_p value at the mean temperature and (iii) by using the mean C_p value.
$C_p = 6.85 + 8.533 \times 10^{-3}T - 2.475 \times 10^{-6}T^2$ kcal $kmol^{-1}K^{-1}$.

7.3 Calculate the heat required to raise the temperature of 28 kg of carbon monoxide from 100°C to 1000°C. Perform the calculation in the following ways: (i) by integrating the expression for C_p, and (ii) by using the mean heat capacity value. C_p for carbon monoxide is given by the expression
$C_p = 6.35 + 1.811 \times 10^{-3}T - 0.267 \times 10^{-6}T^2$, where C_p is in kcal $kmol^{-1}K^{-1}$ and T is in K.

7.4 The gases leaving a combustion chamber have the following composition: carbon dioxide 7.8%, carbon monoxide 0.6%, oxygen 3.4%, water vapour 15.6% and nitrogen 72.6%. Calculate the heat removed if the gases are cooled from 800°C to 200°C.

Table 7.15 Mean heat capacity of the gases in J $mol^{-1}K^{-1}$ based on a reference temperature of 0°C

	200°C	800°C
N_2	29.24	30.77
O_2	29.95	32.52
CO_2	40.15	47.94
CO	29.52	31.10
Water vapour	34.12	37.38

7.5 Estimate the specific enthalpy of ethyl alcohol in kJ kg^{-1} at 1 bar and 200°C, taking the datum temperature as 0°C.

C_p of liquid at 0°C = 24.65 cal mol^{-1}°C^{-1}, and at 80°C = 37.96 cal mol^{-1}°C^{-1}
C_p of gas = $14.66 + 3.76 \times 10^{-2}T - 2.09 \times 10^{-5}T^2$ cal mol^{-1}°C^{-1}
Boiling point of ethanol at 1 bar = 78°C
Latent heat of vapourisation = 9.22 kcal mol^{-1}

7.6 Calculate the standard heat of reaction of the following reaction:

$4NH_3(g) + 5O_2(g) \rightarrow 4NO(g) + 6H_2O(g)$

Standard enthalpy of formation in kJ mol^{-1} for gaseous ammonia, nitric oxide and water vapour is −46.2, 90.3 and −241.6 respectively.

7.7 Calculate the standard heat of reaction of the following reaction:

$C_6H_6(g) + 3H_2(g) \rightarrow C_6H_{12}(g)$

Standard enthalpy of combustion in kJ mol^{-1} of gaseous benzene, hydrogen and cyclohexane is −3 287.4, −285.6 and −3 949.2 respectively.

7.8 Calculate the standard heat of reaction of the following reactions:

(i) Methane–steam reaction to give carbon dioxide and hydrogen;
(ii) Calcium carbide reacting with water to give calcium oxide and acetylene.

The standard heats of formation in kJ mol^{-1} for methane, calcium carbide, acetylene, water vapour, calcium oxide, carbon dioxide and water are −72, −60, +2 168, −232, −608 and −376 respectively.

7.9 The analysis of 15 000 m^3 of a gas at standard conditions is as follows: sulphur dioxide 10%, oxygen 12%, and nitrogen 78%. How much heat must be added to raise the temperature from 30°C to 425°C?

Table 7.16 Mean molar heat capacity in cal mol^{-1}°C^{-1} between 0°C and T°C

Gas temperature	30°C	425°C
SO_2	10.0	11.0
O_2	6.96	7.32
N_2	6.8	7.72

7.10 Calculate the standard heat of reaction of $CO + H_2O \rightarrow CO_2 + H_2$.

Table 7.17 Standard heat of formation

Component	CO	H_2	CO_2
ΔH_f in kJ	−110.6	−242	−394

7.11 Carbon monoxide at 100°C and atmospheric pressure is burnt with 100% excess dry air at 200°C. The products of combustion leave the reaction chamber at 800°C. Calculate the heat evolved in the reaction chamber in kcal kmol^{-1} of CO burnt, assuming complete combustion. Standard heat of combustion of carbon monoxide is = −67.636 kcal kmol^{-1}. Mean heat capacities of carbon monoxide, air, carbon dioxide, oxygen and nitrogen are 7.02, 7.23, 11.92, 7.94 and 7.06 kcal kmol^{-1}K^{-1}.respectively.

7.12 Sulphur dioxide gas is oxidised with 150% excess air with 60% conversion to sulphur trioxide. The gases enter the converter at 673 K and leave at 723 K. How many joules are absorbed in the heat exchanger of the converter per kilomole of sulphur dioxide fed? Standard heat of reaction is −98.9 kJ mol^{-1} for sulphur trioxide. Mean heat capacities of sulphur dioxide, oxygen, nitrogen and sulphur trioxide are 11.08, 7.46, 7.13 and 15.52 cal mol^{-1}K^{-1} respectively.

7.13 Calculate the theoretical flame temperature of a gas having 30% carbon monoxide and 70% nitrogen when burnt with 150% excess air; both air and gas being at 298 K. Standard heat of formation of CO_2 is −94 052 cal mol^{-1} and that of CO is −26 412 cal mol^{-1}. Assume mean C_p values of carbon dioxide, oxygen and nitrogen to be 12.1, 7.9 and 7.5 cal mol^{-1}K^{-1} respectively.

7.14 Calculate the theoretical flame temperature for methane burnt with 100% excess air. The standard heat of reaction is −802 kJ per mol of methane. Determine the final temperature attained by the gaseous products of combustion if the combustion is adiabatic and complete. 20% of the heat is lost to surroundings. Mean specific heat in cal mol^{-1}K^{-1} for oxygen, nitrogen, water vapour and carbon dioxide is 31.9, 32.15, 40.18 and 51.8 respectively.

7.15 Iron pyrite ore contains 85% iron pyrite and the rest gangue. It is roasted with 200% excess air to get sulphur dioxide. All the gangue and ferric oxide formed end up in the solid waste which contains 4% unreacted iron pyrite. Determine the standard heat of reaction in kJ kg^{-1} of ore roasted. Heats of formation in kJ kmol^{-1} of iron pyrite, ferric oxide and sulphur dioxide are −1.78, −88 and −297 respectively.

7.16 Coal is burned to a gas of the following composition: carbon dioxide 9.2%, carbon monoxide 1.5%, oxygen 7.3% and the rest nitrogen. What is the enthalpy change for this gas between the bottom of the stack (280°C) and the top of the stack (90°C)?

$$C_p\ (N_2) = 6.895 + 0.7624 \times 10^{-3}T - 0.7 \times 10^{-7}T^2 \text{ kcal kmol}^{-1}\text{K}^{-1}.$$

$$C_p\ (O_2) = 7.104 + 0.7851 \times 10^{-3}T - 0.55 \times 10^{-7}T^2 \text{ kcal kmol}^{-1}\text{K}^{-1}.$$

$$C_p\ (CO) = 6.865 + 0.8026 \times 10^{-3}T - 0.7367 \times 10^{-7}T^2 \text{ kcal kmol}^{-1}\text{K}^{-1}.$$

$$C_p\ (CO_2) = 8.448 + 0.5757 \times 10^{-3}T - 0.216 \times 10^{-7}T^2 \text{ kcal kmol}^{-1}\text{K}^{-1}.$$

7.17 Calculate the theoretical flame temperature of a gas having 20% carbon monoxide and 80% nitrogen, when burnt with 150% excess air, both air and gas being at 298 K. Standard heat of formation of carbon dioxide and carbon monoxide in cal mol^{-1} are −94 052 and −26 412 respectively. Assume mean C_p values of carbon dioxide, oxygen and nitrogen to be 12.0, 7.9 and 7.2 cal mol^{-10}C^{-1} respectively.

8

Fuels and Combustion

A *fuel* is a combustible substance which on burning releases a large amount of heat energy.

$$\text{Fuel} + \text{oxygen} \rightarrow \text{products of combustion} + \text{heat energy}$$

The combustibles in the fuel are carbon, hydrogen and sulphur. The products of combustion of a fuel are known as *flue gases*.

CALORIFIC VALUE OF FUEL

The *calorific value* of a fuel is the standard heat of combustion of the fuel. It is defined as the total heat produced when unit mass of the fuel is completely burnt with pure oxygen. The calorific value is also known as the heating value.

When a fuel is burnt, hydrogen combines with oxygen and gets converted to water.

$$H_2 + 1/2\,O_2 \rightarrow H_2O$$

1 mol or 2 g of hydrogen combine with 0.5 mol or 16 g of oxygen to give 1 mol or 18 g of water. Hence the weight of water formed during the combustion of a fuel is nine times the weight of hydrogen present in the fuel.

When water vapour is present in the flue gas, the latent heat of vapourisation is not available as heat energy of the fuel. The calorific value of the fuel determined, considering that water is present in the vapour form, is known as the *net calorific value* (NCV), *lower heating value* (LHV) or *net heating value* (NHV). If the vapours are condensed, the latent heat of vapourisation (λ_v) can be recovered. If this heat is added to the net calorific value, *gross calorific value* (GCV), *gross heating value* (GHV) or *total heating value* (THV) is obtained.

Usually the calorific value of fuels are reported at 25°C and expressed in kJ or kcal kg^{-1}, or kJ or kcal m^{-3}. When the calorific value is reported it should be specified whether it is NCV or GCV. If no mention is made, the reported value should be taken as GCV.

The calorific value of a fuel is determined using a calorimeter of known weight, in which a known weight of the fuel is burnt with pure oxygen and the heat liberated is absorbed in water of known weight. From the rise in temperature of the water and the calorimeter, GCV of the fuel is calculated.

$$\text{NCV} = \text{GCV} - \text{weight fraction of hydrogen} \times 9 \times \lambda_v \text{ kJ kg}^{-1} \text{ at 298 K},$$

where λ_v, the latent heat of vapourisation of water at 298 K is 2 442.5 kJ kg^{-1}.

ANALYSIS OF FUEL

The analysis of a fuel is of two types: proximate analysis and ultimate analysis. Proximate analysis determines the moisture, volatile matter, ash and fixed carbon in the fuel. Ultimate analysis is the elemental analysis of the fuel. Here the fuel is analysed for its hydrogen, carbon, nitrogen, sulphur, oxygen and ash content.

Proximate analysis

Moisture: The moisture content of a fuel is the loss in weight of one gram of sample when heated or dried in an oven at 381 ± 2 K for 1 hour. Loss is expressed as percentage.

Volatile matter: The volatile content of a fuel is the loss in weight of 1 g of sample heated at 1 223 K for 7 minutes, minus the moisture content. Loss is expressed as percentage.

Ash: The ash content of a fuel is the weight of residue remaining after all the combustible material is burnt when one gram of sample is heated at 973–1 023 K. This also is expressed as percentage.

Fixed carbon: Fixed carbon content of the fuel is the remainder when the sum of the moisture, volatile matter and ash contents are subtracted from 100.

Ultimate analysis

Carbon and Hydrogen: A known weight (X) of the sample is burnt in air. Carbon and hydrogen are oxidised to carbon dioxide and water vapour respectively.

$$C + O_2 \rightarrow CO_2$$

12 g of carbon on burning give 44 g of carbon dioxide.

$$H_2 + 1/2\, O_2 \rightarrow H_2O$$

2 g of hydrogen on burning give 18 g of water vapour.

The gases are first passed through a bulb containing aqueous potassium hydroxide solution and then through a tube containing anhydrous calcium chloride. The bulbs are weighed before and after the passage of the gas.

$$\text{Percentage of C} = \frac{\text{increase in weight of potassium hydroxide bulb} \times 12 \times 100}{44X}$$

$$\text{Percentage of H} = \frac{\text{increase in weight of calcium chloride tube} \times 2 \times 100}{18X}$$

Nitrogen: A known weight of the sample is refluxed with concentrated sulphuric acid in the presence of potassium sulphate catalyst. When the solution becomes clear, excess of potassium hydroxide solution is added and boiled. The resulting ammonia gas is absorbed in standard sulphuric acid of known strength.

The reactions occurring are

$$2N + 6H + H_2SO_4 \rightarrow (NH_4)_2SO_4$$

$$(NH_4)_2SO_4 + 2KOH \rightarrow 2NH_3 + K_2SO_4 + 2H_2O$$

$$2NH_3 + H_2SO_4 \rightarrow (NH_4)_2SO_4$$

1000 ml of 1 N sulphuric acid is equivalent to 14 g of nitrogen.

$$\text{Percentage of } N_2 = \frac{\text{volume of acid used in absorption} \times \text{strength} \times 14 \times 100}{1000X}$$

Sulphur: A known mass of the sample is burnt in a calorimeter. The residue is dissolved in distilled water and treated with barium chloride solution. Barium sulphate precipitates out and the precipitate is filtered and weighed. 233 g of barium sulphate contain 32 g of sulphur.

$$\text{Percentage of S} = \frac{\text{weight of barium sulphate} \times 32 \times 100}{233X}$$

Ash: The ash content of the sample is determined following the method described in the case of proximate analysis.

Oxygen: The oxygen content of the fuel is the remainder after the sum of the carbon, hydrogen, nitrogen, sulphur and ash content are subtracted from 100.

$$\text{Percentage of } O_2 = C - H - S - N - \text{Ash}$$

Theoretical Calorific Value

The calorific value of a fuel can be theoretically estimated from the weight fractions of combustibles and their heating values. Gross calorific value of the fuel is the algebraic sum of the heating values of the combustibles present in the fuel. The heating value of carbon, hydrogen and sulphur are 33 950, 144 200 and 9 400 kJ kg^{-1} respectively.

$$\text{Gross calorific value of the fuel} = 33\,950\ C + 144\,200\left[H - \left(\frac{O}{8}\right)\right] + 9\,400\ S \text{ kJ kg}^{-1}$$

where C, O, H and S are the weight fractions of carbon, oxygen, hydrogen and sulphur in the fuel.

ILLUSTRATIONS

8A A sample of fuel oil has a carbon to hydrogen ratio of 9.33 by weight, and also contains sulphur to the extent of 1.37% by weight. The gross calorific value of the fuel is measured as 41 785 kJ kg^{-1} at 298 K. Calculate its net calorific value at 298 K.

Solution

Basis: 1 kg of the fuel

The fuel has a carbon to hydrogen ratio of 9.33 by weight and also contains sulphur to the extent of 1.37% by weight.

Carbon and hydrogen in the fuel = 100 – 1.37 = 98.63%

Out of 10.33 parts of carbon and hydrogen, one part is hydrogen.

$$\text{Hydrogen in the fuel} = \frac{98.63 \times 1}{10.33} = 9.52\%$$

$$\text{Hydrogen in 1 kg of fuel} = \frac{9.52 \times 1000}{100} = 95.2 \text{ g}$$

Hydrogen on combustion forms water as per the reaction

$$H_2 + 1/2\, O_2 \rightarrow H_2O$$

1 g of hydrogen gives 9 g of water.

Water formed on combustion = 95.2 × 9 = 856.8 g = 0.856 8 kg

λ_v at 298 K is 2 442.5 kJ kg^{-1}.

Net calorific value = gross calorific value – weight of water formed × λ_v kJ kg^{-1} at 298 K

= 41 785 – 0.8 568 × 2 442.5 = 39 685 kJ kg^{-1}

8B Gross calorific value of (i) gaseous *n*-propane, and (ii) liquid acetone at 298 K are 2 219.2 kJ mol^{-1} and 1 821.3 kJ mol^{-1} respectively. Find their net calorific value in kJ kg^{-1}.

Solution

Basis: 1 mol of fuel burnt

(i) The reaction is

$$C_3H_8 + 5O_2 \rightarrow 3CO_2 + 4H_2O$$

Each mole of fuel, on combustion gives 4 mol or 72 g or 0.072 kg of water.

Net calorific value = gross calorific value – weight of water formed × λ_v kJ kg^{-1} at 298 K.
λ_v at 298 K is 2 442.5 kJ kg^{-1}.

Net calorific value $= 2\,219.2 - 0.072 \times 2\,442.5 = 2\,043.3$ kJ mol^{-1}

Molecular weight of propane is 72. 1 mol of propane weighs 72 g.

$$\text{Net calorific value} = \frac{2\,043.3 \times 1000}{72} = 46\,438.6 \text{ kJ kg}^{-1}$$

(ii) The reaction is

$$C_3H_6O + 4O_2 \rightarrow 3CO_2 + 3H_2O$$

Each mole of fuel on combustion gives 3 mol or 54 g or 0.054 kg of water.

Net calorific value = gross calorific value – weight of water formed × λ_v kJ kg^{-1} at 298 K.
λ_v at 298 K is 2 442.5 kJ kg^{-1}.

Net calorific value $= 1\,821.3 - 0.054 \times 2\,442.5 = 1\,689.4$ kJ mol^{-1}

Molecular weight of acetone is 54. One mole of acetone weighs 54 g.

$$\text{Net calorific value} = \frac{1\,689.4 \times 1000}{54} = 29\,127.6 \text{ kJ kg}^{-1}$$

COMBUSTION OF FUEL

Theoretical oxygen demand (ThOD) is the minimum amount of oxygen required to burn the fuel completely so that the combustibles in it namely carbon, hydrogen and sulphur are converted to carbon dioxide, water vapour and sulphur dioxide respectively.

$$C + O_2 \rightarrow CO_2$$

$$H_2 + 1/2\, O_2 \rightarrow H_2O$$

$$S + O_2 \rightarrow SO_2$$

Each atom of carbon and sulphur requires 1 mol of oxygen. 1 mol of hydrogen requires 0.5 mol of oxygen.

ThOD of the fuel = sum of the ThOD of all combustibles present in the fuel
= ThOD of C + ThOD of H_2 + ThOD of S

The source of oxygen is air; mole fraction of oxygen in air is 0.21.

$$\text{Theoretical air demand in moles} = \frac{\text{theoretical oxygen demand in moles}}{0.21}$$

In practice, theoretical air is not sufficient for complete combustion of the fuel. Excess of oxygen or air is a must for complete combustion.

$$\text{Percentage of excess air} = \frac{\text{Actual air supplied} - \text{theoretical air demand} \times 100}{\text{Theoretical air}}$$

When a fuel is partially burnt, the flue gases will contain carbon monoxide in addition to carbon dioxide. Carbon monoxide can consume oxygen for its conversion to carbon dioxide. So the excess oxygen present in flue gases containing carbon monoxide is termed as apparent excess. To find the actual excess oxygen, the ThOD of the carbon monoxide (0.5 mol per mole of carbon monoxide) has to be subtracted from the apparent excess oxygen.

The excess air depends on the fuel type. Excess air requirement is 50% for solid fuels, 30% for liquid fuels and 20% for gaseous fuels. Excess air reduces the flame temperature and increases heat losses through flue gases.

Some facts about air are given below:

Average molecular weight of air = 29
Volume or mole per cent of oxygen in air = 21 vol% or 21 mol%
Weight per cent of oxygen in air = 23%
1 kmol of air at NTP occupies 22.4 m^3.

In air, 1 mol oxygen is accompanied by 3.76 mol nitrogen.

ORSAT ANALYSIS

An Orsat analyser is used for the analysis of flue gases to determine the volume per cent of carbon dioxide, oxygen and carbon monoxide in it. This apparatus consists of a gas measuring burette and three absorption pipettes, each containing an aqueous solution, which absorbs only one of the gases. Alkaline pyrogallol is used for absorbing oxygen, aqueous potassium hydroxide solution for carbon dioxide and sulphur dioxide, and ammoniacal cuprous chloride for carbon monoxide.

100 ml of the flue gas sample is measured accurately in the gas burette. Carbon dioxide in the sample is first absorbed by passing the gas sample back and forth between the absorption pipette and measuring burette, until no further absorption takes place. The shrinkage in gas volume directly gives the percentage of carbon dioxide in the sample. The same procedure is repeated for finding oxygen, and finally the carbon monoxide using the separate absorption pipettes provided for each gas. The gas remaining unabsorbed gives the percentage of nitrogen in the sample.

The gas in the measuring burette is in contact with water. All absorbing solutions are aqueous solutions. So the gas becomes rapidly saturated with water vapour at the temperature of the instrument. The partial pressure of water vapour is the same at all readings and the percentage of gas absorbed is also the same, as though the gas is dry from the beginning to the end. So Orsat analysis gives the composition of the flue gas on a dry basis. As sulphur dioxide is also absorbed by alkali along with carbon dioxide, carbon dioxide and sulphur dioxide are expressed as carbon dioxide only, as it is the major constituent.

ILLUSTRATIONS

8C Determine the volume of air needed for complete combustion of 1 m^3 of producer gas having the following composition by volume: hydrogen 30%, carbon monoxide 10%, methane 4% and nitrogen 56%.

Solution

Basis: 1 m^3 of the producer gas

The combustion reactions are:

$$H_2 + 1/2\ O_2 \rightarrow H_2O$$

1 mol or 1 volume of hydrogen requires 1/2 volume of oxygen for combustion.

$$CH_4 + 2O_2 \rightarrow CO_2 + 2H_2O$$

1 mol or 1 volume of methane requires 2 volumes of oxygen for combustion.

$$CO + 1/2\ O_2 \rightarrow CO_2$$

1 mol or 1 volume of carbon monoxide requires 1/2 volume of oxygen for combustion.
The ThOD of the fuel is given in Table 8.1.

Table 8.1 Theoretical oxygen demand of the fuel

Constituent	Volume (m^3)	ThOD (m^3)
H_2	0.3	0.15
H_2O	0.1	0.05
CH_4	0.04	0.08
ThOD of producer gas		0.28

Air contains 21% by volume of oxygen.

Volume of air required per cubic metre of fuel $= \dfrac{0.28 \times 100}{21} = 1.333\ m^3$, at the same pressure and temperature at which fuel is fed.

8D A furnace uses coke containing 80% carbon, 0.5% hydrogen and 19.5% ash by weight. It operates with 50% excess air. The solid refuse (ash) contains 2% unburnt carbon. 95% of carbon burnt forms carbon dioxide and the balance carbon monoxide. Calculate (i) Orsat analysis of flue gas, (ii) the weight of ash produced per 100 kg coke burnt, and (iii) the weight of carbon lost per 100 kg coke burnt

Solution

Basis: 100 kg of coke burnt

Coke contains 19.5% ash. So ash in 100 kg coke = 19.5 kg. This forms 98% of solid refuse from the furnace and the balance of 2% is unburnt carbon.

Weight of solid refuse from the furnace $= \frac{19.5 \times 100}{98} = 19.9$ kg

Weight of ash in solid refuse = 19.5 kg

Weight of ash produced per 100 kg of coke burnt = 19.5 kg

Weight of unburnt carbon in refuse = 0.4 kg

Weight of carbon lost per 100 kg of coke burnt = 0.4 kg

The combustibles in the fuel are carbon and hydrogen. The combustion reactions are

$$C + O_2 \rightarrow CO_2$$

$$H_2 + 1/2\, O_2 \rightarrow H_2O$$

1 mol of carbon consumes 1 mol of oxygen to give 1 mol of carbon dioxide. 1 mol of hydrogen consumes 0.5 mol of oxygen to give 1 mole of water. The ThOD of the fuel is given in Table 8.2.

Table 8.2 ThOD of the fuel

Constituent	Weight (kg)	Molecular weight	Amount (kmol)	ThOD
C	80	12	6.667	6.667
H_2	0.5	2	0.125	0.125
				6.792

50% excess oxygen is supplied.

$$\text{Oxygen supplied} = \frac{6.792 \times 150}{100} = 10.19 \text{ kmol}$$

The source of oxygen is air. So oxygen is accompanied by nitrogen. Air contains 21mol% oxygen and 79mol% nitrogen.

$$\text{Nitrogen accompanying oxygen} = \frac{10.19 \times 79}{21} = 38.33 \text{ kmol}$$

Carbon burnt = 80 − 0.4 = 79.6 kg = 6.63 kmol

95% of the carbon burnt forms carbon dioxide and the balance carbon monoxide.

Carbon dioxide formed $= 6.63 \times 0.95 = 6.30$ kmol

Carbon monoxide formed $= 6.63 - 6.30 = 0.33$ kmol

1 mol of carbon dioxide requires 1 mol of oxygen and 1 mol of carbon monoxide requires 0.5 mol of oxygen.

Oxygen consumed by carbon $= 6.30 + (0.33 \div 2) = 6.465$ kmol

1 mol of hydrogen requires 0.5 mol of oxygen.

Oxygen consumed by hydrogen = 0.125 kmol

Total oxygen consumed $= 6.465 + 0.125 = 6.59$ kmol

Excess of oxygen = oxygen supplied − oxygen consumed

$= 10.19 - 6.59 = 3.60$ kmol

Orsat analysis is on water vapour-free basis and is given in Table 8.3.

Table 8.3 Orsat analysis of the products of combustion

Constituent	Amount (kmol)	Volume per cent
CO_2	6.30	12.97
CO	0.33	0.68
O_2	3.60	7.42
N_2	38.33	78.93
	48.50	

8E The composition of a coal sample by weight is found to be carbon 76%, hydrogen 5.2%, oxygen 12.8%, nitrogen 2.7% and sulphur 1.2%. The remaining is ash. Calculate the minimum (i) weight and (ii) volume at NTP for complete combustion of 1 kg of coal, and (iii) percentage composition by weight of dry products, if 50% excess air is supplied.

Solution

Basis: 1 kg of coal

1 kg coal contains carbon 760 gm, hydrogen 52 gm, sulphur 12 gm and oxygen 128 g.
The combustion reactions are

$$C + O_2 \rightarrow CO_2$$

$$H_2 + 1/2\, O_2 \rightarrow H_2O$$

$$S + O_2 \rightarrow SO_2$$

12 g of carbon consume 32 g of oxygen to give 44 g of carbon dioxide.
2 g of hydrogen consume 16 g of oxygen to give 18 g of water.
32 g of sulphur consume 32 g of oxygen to give 64 g of sulphur dioxide.

The ThOD is calculated by summing up the oxygen requirement for each of the combustibles in the fuel and subtracting from that the oxygen present in the fuel.

$$\text{ThOD} = \frac{760 \times 32}{12} + \frac{56 \times 16}{2} + \frac{12 \times 32}{32} - 128 = 2\,326.67\text{ g} = 2.327\text{ kg}$$

Air contains 23% by weight of oxygen and 77% by weight of nitrogen. The average molecular weight of air is 29. 1 kmol of air at NTP occupies a volume of 22.4 m^3.

$$\text{Minimum air required} = \frac{2.327 \times 100}{23} = 10.116\text{ kg}$$

$$= 0.349\text{ kmol}$$

$$= 7.814\text{ Nm}^3$$

50% excess air is supplied in this process.

$$\text{Air supplied} = \frac{10.116 \times 150}{100} = 15.249\text{ kg} = 15\,249\text{ g}$$

$$\text{Amount of oxygen in air} = \frac{15\,249 \times 23}{100} = 3\,507\text{ g}$$

$$\text{Amount of nitrogen in air} = \frac{15\,249 \times 77}{100} = 11\,742\text{ g}$$

$$\text{Amount of nitrogen in coal} = 27\text{ g}$$

$$\text{Total nitrogen} = 11\,769\text{ g}$$

Excess of oxygen = oxygen supplied ThOD = 3 507 − 2 327 = 1 180 g

The composition of dry products of combustion is given in Table 8.4.

Table 8.4 Composition of dry products of combustion

Constituent		Weight (g)	Weight per cent
CO_2	$\frac{760 \times 44}{12}$	2 787	17.68
SO_2	$\frac{12 \times 64}{32}$	24	0.15
N_2		11 769	74.68
O_2		1 180	7.49
		15 760 g	

8F Butane is burnt with 80% of theoretical air. Calculate the analysis of gases leaving, assuming that all the hydrogen present in butane is converted into water.

Solution

Basis: 1 mol of butane burnt

The reaction is

$$C_4H_{10} + 13/2\ O_2 \rightarrow 4CO_2 + 5H_2O$$

$$\text{ThOD} = 6.5\ \text{mol}$$

But only 80% oxygen is supplied.

Oxygen supplied $= 6.5 \times 0.8 = 5.2$ mol

The source of oxygen is air. Air contains 21mol% of oxygen and 79mol% of nitrogen.

Amount of nitrogen accompanying oxygen in air $= \dfrac{5.2 \times 79}{21} = 19.56$ mol

Insufficient oxygen is supplied. We have been asked to assume that all hydrogen is burnt.

Hydrogen in 1 mol of butane = 5 mol

1 mol of hydrogen consumes 0.5 mol of oxygen to give 1 mol of water, as per the reaction,

$$H_2 + 1/2\ O_2 \rightarrow H_2O$$

Oxygen required for the complete combustion of hydrogen in butane $= 2.5$ mol

Water formed $= 5$ mol

Oxygen available for the combustion of carbon $= 5.2 - 2.5 = 2.7$ mol

1 mol of carbon consumes 1 mol of oxygen to give 0.5 mol of carbon dioxide as per the reaction,

$$C + O_2 \rightarrow CO_2$$

Carbon dioxide formed $= 2.7$ mol

Table 8.5 gives the analysis of the products of combustion.

Table 8.5 Analysis of the products of combustion

Constituent	Moles (mol)	Mole per cent
CO_2	2.70	9.90
Water vapour	5.00	18.34
N_2	19.56	71.76
	27.26	

8G Natural gas containing 82.4% methane, 6.8% ethane, 3.78% propane and the balance butane is burnt with 30% excess air. Calculate (i) flue gas analysis, and (ii) volume of flue gases per cubicmetre of gas burnt.

Solution

Basis: 1 mol of natural gas

The composition is given in mole per cent. As 1 mol of natural gas has been selected as the basis, the composition expressed as mole fraction gives the number of moles of each constituent in the mixture. The amount of each constituent is given below:

Amount of methane = 0.824 0 mol

Amount of ethane = 0.068 0 mol

Amount of propane = 0.037 8 mol

Amount of butane = 0.070 2 mol

1 mol of methane consumes 2 mol of oxygen to give 1 mol of carbon dioxide and 2 mol of water vapour as per the reaction,

$$CH_4 + 2O_2 \rightarrow CO_2 + 2H_2O$$

1 mol of ethane consumes 3.5 mol of oxygen to give 2 mol of carbon dioxide and 3 mol of water vapour as per the reaction,

$$C_2H_6 + 7/2\ O_2 \rightarrow 2CO_2 + 3H_2O$$

1 mol of propane consumes 5 mol of oxygen to give 3 mol of carbon dioxide and 4 mol of water vapour as per the reaction,

$$C_3H_8 + 5O_2 \rightarrow 3CO_2 + 4H_2O$$

1 mol of butane consumes 6.5 mol of oxygen to give 4 mol of carbon dioxide and 5 mol of water vapour as per the reaction,

$$C_4H_{10} + 13/2\ O_2 \rightarrow 4CO_2 + 5H_2O$$

Using the above information, ThOD can be calculated, which is the sum of the oxygen requirement of all the constituents.

$$\text{ThOD} = (2 \times 0.824) + (3.5 \times 0.068) + (5 \times 0.0378) + (6.5 \times 0.0702)$$

$$= 2.5313 \text{ mol}$$

30% excess oxygen is supplied.

$$\text{Oxygen supplied} = \frac{2.5313 \times 130}{100} = 3.291 \text{ mol}$$

The source of oxygen is air. Air contains 21mol% oxygen and 79mol% nitrogen.

$$\text{Nitrogen accompanying oxygen in air} = \frac{3.291 \times 79}{21} = 12.379 \text{ mol}$$

Amount of carbon dioxide and water vapour formed can be calculated from the amounts of the constituents of natural gas, and their combustion reactions explained above, as given in Table 8.6.

Table 8.6 Calculation of products of combustion

Constituent	Moles (mol)	Carbon dioxide formed (mol)	Water vapour formed (mol)
CH_4	0.8240	$0.8240 \times 1 = 0.824$	$0.8240 \times 2 = 1.648$
C_2H_6	0.0680	$0.0680 \times 2 = 0.136$	$0.0680 \times 3 = 0.204$
C_3H_9	0.0378	$0.0378 \times 3 = 0.1134$	$0.0378 \times 4 = 0.1512$
C_4H_{10}	0.0702	$0.0702 \times 4 = 0.2808$	$0.0702 \times 5 = 0.3510$
		1.3536	2.3542

Analysis of the products of combustion is given in Table 8.7.

Table 8.7 Analysis of the products of combustion

Constituent	Amount (mol)	Mole per cent
CO_2	1.3536	8.04
Water vapour	2.3542	13.97
N_2	12.379	73.48
O_2	0.7594	4.51
	16.8462	

1 mol of fuel produces 16.8462 mol of flue gas.

For gases, we know that the composition by mole and volume are analogous.

Volume of flue gas produced per cubicmetre of fuel gas burnt = 16.8462 m^3 (assuming the fuel enters and flue gases leave the furnace at the same temperature and pressure).

8H Analysis of a coal sample gave the following result: Carbon 61.5%, hydrogen 3.5%, sulphur 0.4%, ash 14.2%, nitrogen 1.8% and oxygen 18.6%. Calculate (i) the theoretical air requirement, and (ii) the actual as well as Orsat analysis of the flue gases if the coal is burnt with 60% excess air.

Solution

Basis: 100 kg of coal

The combustion reactions are

$$C + O_2 \rightarrow CO_2$$

$$H_2 + 1/2\ O_2 \rightarrow H_2O$$

$$S + O_2 \rightarrow SO_2$$

1 mol of carbon consumes 1 mol of oxygen to give 1 mol of carbon dioxide.
1 mol of hydrogen consumes 0.5 mol of oxygen to give 1 mol of water vapour.
1 mol of sulphur consumes 1 mol of oxygen to give 1 mol of sulphur dioxide.

ThOD is calculated by summing up the oxygen requirement for each combustible in the fuel and subtracting from that the oxygen present in the fuel, as given in Table 8.8.

Table 8.8 ThOD of the fuel

Constituent	Weight (kg)	Amount (kmol)	Oxygen demand
C	61.5	5.125	+5.125
H_2	3.5	1.75	+0.875
S	0.4	0.012 5	+0.012 5
N_2	1.8	0.064 3	–
O_2	18.6	0.581 3	−0.581 3
			5.431 2

The source of oxygen is air. Air contains 21mol% oxygen and 79mol% nitrogen.

$$\text{Theoretical air demand} = \frac{5.431\,2}{0.21} = 25.86\ \text{kmol per 100 kg of coal}$$

60% excess air is supplied.

$$\text{Air supplied} = \frac{25.86 \times 160}{100} = 41.376\ \text{kmol}$$

$$\text{Amount of oxygen in air} = \frac{41.376 \times 21}{100} = 8.689\ \text{kmol}$$

$$\text{Amount of nitrogen in air} = \frac{41.376 \times 79}{100} = 32.687\ \text{kmol}$$

Amount of nitrogen in coal = 0.0643 kmol

Total nitrogen present = 32.751 kmol

Excess oxygen = oxygen supplied − ThOD

$= 8.689 - 5.431 = 3.258$ kmol

1 mol of carbon on complete combustion gives 1 mol of carbon dioxide.
1 mol of hydrogen on complete combustion gives 1 mol of water vapour.
1 mol of sulphur on complete combustion gives 1 mol of sulphur dioxide.

Carbon dioxide formed = 5.125 kmol

Water vapour formed = 1.75 kmol

Sulphur dioxide formed = 0.012 5 kmol

Table 8.9 gives the actual flue gas analysis.

Table 8.9 Actual flue gas analysis

Constituent	Amount (kmol)	Volume per cent
CO_2	5.125	11.94
Water vapour	1.75	4.08
SO_2	0.012 5	0.03
N_2	32.751	76.35
O_2	3.258	7.60
	42.896 5	

In an Orsat analyser, carbon dioxide and sulphur dioxide are read as carbon dioxide, as potassium hydroxide solution absorbs sulphur dioxide along with carbon dioxide. Table 8.10 gives the Orsat analysis of the flue gas.

Table 8.10 Orsat analysis of the flue gas

Constituent	Kilomoles (kmol)	Volume per cent
CO_2	5.137 5	12.48
N_2	32.751	79.60
O_2	3.258	7.92
	41.146 5	

8I A gas containing 75% methane and 25% oxygen is burnt with 100% excess air. 80% methane goes to carbon dioxide, 10% to carbon monoxide and the rest remains unburnt. Calculate the exit gas composition.

Solution

Basis: 100 moles of fuel gas

Amount of methane in the fuel = 75 mol

Amount of oxygen in the fuel = 25 mol

The combustion reaction is

$$CH_4 + 2O_2 \rightarrow CO_2 + 2H_2O$$

ThOD is calculated by summing up the oxygen requirement for each combustible in the fuel and subtracting from that the oxygen present in the fuel.

ThOD = 75 × 2 – 25 = 125 mol

100% excess oxygen is supplied.

Oxygen supplied $= \dfrac{125 \times 200}{100} = 250$ mol

Amount of oxygen in the fuel = 25 mol

Total oxygen in the input = 275 mol

The source of external oxygen is air. Air contains 21 mol% oxygen and 79 mol% nitrogen.

Amount of nitrogen from air $= \dfrac{250 \times 79}{21} = 940$ mol

80% of methane is burnt to carbon dioxide, as per the reaction,

$$CH_4 + 2O_2 \rightarrow CO_2 + 2H_2O$$

Methane burnt to carbon dioxide $= \dfrac{75 \times 80}{100} = 60$ mol

Oxygen consumed = 60 × 2 = 120 mol

Carbon dioxide formed = 60 × 1 = 60 mol

Water vapour formed = 60 × 2 = 120 mol

10% of methane is burnt to carbon monoxide, as per the reaction,

$$CH_4 + 1\,1/2\ O_2 \rightarrow CO + 2H_2O.$$

Methane burnt to carbon monoxide = 75 × 0.1 = 7.5 mol

Oxygen consumed = 7.5 × 1.5 = 11.25 mol

Carbon monoxide formed = 7.5 × 1 = 7.5 mol

Water vapour formed = 7.5 × 2 = 15 mol

The remaining 10% of methane is unburnt.

Unburnt methane $= 75 \times 0.1 = 7.5$ mol

Total oxygen consumed $= 120 + 11.25 = 131.25$ mol

Excess of oxygen = oxygen supplied − oxygen consumed

$= 275 - 131.25 = 143.75$ mol

Total water vapour formed $= 120 + 15 = 135$ mol

Table 8.11 gives the exit gas composition.

Table 8.11 Exit gas composition

Constituent	Amount (mol)	Mole per cent
CO_2	60.00	4.64
CO	7.50	0.58
CH_4	7.50	0.58
Water vapour	135.00	10.44
N_2	940.00	72.66
O_2	143.75	11.10
	1 293.75	

8J Coke containing 85% carbon and 15% ash is burnt in a furnace. Analysis of flue gas from the chimney shows carbon dioxide 12% and carbon monoxide 3% the balance being oxygen and nitrogen. Find (i) the percentage of excess air, (ii) the weight of flue gas per kilogram of coke burnt, and (iii) the air to fuel ratio in kg^{-1}.

Solution

Basis: 100 kmol of flue gas. We do not have the entire information on the input side; for example, the percentage of excess air supplied is not known. So the problem has to be solved choosing the output as the basis.

Oxides of carbon in 100 kmol of flue gas $= 12 + 3 = 15$ kmol

1 mol carbon gives 1 mol of carbon dioxide on complete combustion or 1 mol of carbon monoxide on partial combustion.

$$C + O_2 \rightarrow CO_2 \text{ or } C + 1/2\, O_2 \rightarrow CO$$

Carbon required for 100 kmol of flue gas $= 15 \times 12 = 180$ kg

Coke contains 85% by weight of carbon.

Coke required for 100 kmol flue gas $= \dfrac{180 \times 100}{85} = 211.76\,\text{kg}$

Unreacted oxygen and inert nitrogen in flue gas $= 85\,\text{kmol}$

Reacted oxygen $=$ oxygen in carbon dioxide $+$ oxygen in carbon monoxide

$= 12.0 + 1.5 = 13.5\,\text{kmol}$

Air supplied $=$ unreacted oxygen and inert nitrogen in flue gas $+$ oxygen reacted

$= 85.0 + 13.5 = 98.5\,\text{kmol} = 2\,856.5\,\text{kg}$

Air contains 21mol% oxygen.

Oxygen supplied as air $= \dfrac{98.5 \times 21}{100} = 20.685\,\text{kmol}$

ThOD $= 15.0\,\text{kmol}$, as the fuel fed contains 15 kmol carbon.

Excess of oxygen $=$ oxygen supplied $-$ ThOD

$= 20.685 - 15.0 = 5.685\,\text{kmol}$

Percentage of excess oxygen $= \dfrac{\text{excess oxygen} \times 100}{\text{ThOD}} = \dfrac{5.685 \times 100}{15} = 39.1\%$

Weight of flue gas $=$ weight of combustible in coke $+$ weight of air supplied

Weight of flue gas formed $= 180 + 2\,856.5 = 3\,036.5\,\text{kg}$

Weight of flue gas per kilogram of coke $= \dfrac{3\,036.5}{211.76} = 14.34\,\text{kg}$

Air to fuel ratio in kilogram $= \dfrac{2\,856.5}{211.76} = 13.49$

8K A liquid fuel having 88wt% carbon and 12wt% hydrogen is burnt to a flue gas of the following composition by volume on dry basis: Carbon dioxide 13.4%, oxygen 3.6% and nitrogen 83.0%. Calculate the amount of flue gas produced per kilogram of fuel. Also find the percentage of excess air.

Solution

Basis: 100 kmol of dry flue gas. We do not have the entire information on the input side; for example, the percentage of excess air supplied is not known. So the problem has to be solved choosing the output as the basis.

Carbon dioxide in 100 kmol of dry flue gas $= 13.4$ kmol

1 mol of carbon gives 1 mol of carbon dioxide on complete combustion.

$$C + O_2 \rightarrow CO_2$$

Carbon required for 100 kmol of flue gas $= 13.4 \times 12 = 160.8$ kg

The fuel contains 88% by weight of carbon.

Fuel required for 100 kmol of flue gas $= \dfrac{160.8 \times 100}{88} = 182.73$ kg

Fuel contains carbon and hydrogen only.

Hydrogen in fuel = weight of fuel − weight of carbon

$= 182.73 - 160.80 = 21.93$ kg $= 10.97$ kmol

1 mol of hydrogen consumes 0.5 mol of oxygen to give 1 mol of water vapour.

$$H_2 + 1/2\, O_2 \rightarrow H_2O$$

Water vapour formed $= 10.97$ kmol

Actual flue gas formed = Amounf of dry flue gas + Amount of water vapour

$= 100 + 10.97 = 110.97$ kmol

Amount of flue gas produced per kilogram of fuel $= \dfrac{110.97}{182.73} = 0.6073 \text{ kmol kg}^{-1}$

For 100 kmol dry flue gas, 182.73 kg fuel containing 13.4 kmol carbon and 10.97 kmol hydrogen is required. 1 mol of carbon consumes 1 mol of oxygen and 1 mol of hydrogen consumes 0.5 mol of oxygen.

ThOD is calculated by summing up the oxygen requirement for each combustible in the fuel.

ThOD $= (13.4 \times 1) + (10.97 \times 0.5) = 18.88$ kmol

Excess of oxygen in 100 kmol of dry flue gas $= 3.6$ kmol

Percentage of excess oxygen $= \dfrac{\text{excess oxygen} \times 100}{\text{ThOD}} = \dfrac{3.6 \times 100}{18.88} = 19.06\%$

8L A hydrocarbon fuel is burnt with excess air. Orsat analysis of the flue gas shows 10.2% carbon dioxide, 1.0% carbon monoxide, 8.4% oxygen and 80.4% nitrogen. Find the percentage of excess air and atomic ratio of hydrogen to carbon in the fuel.

Solution

Basis: 100 mol of dry flue gas. We do not have the entire information on the input side; for example the percentage of excess air supplied is not known. So the problem has to be solved choosing the output as the basis.

Amount of nitrogen in 100 mol of dry flue gas = 80.4 mol

The fuel is a hydrocarbon. The source of nitrogen in the flue gas is the air supplied for combustion. Air contains 21mol% oxygen and 79mol% nitrogen.

$$\text{Amount of oxygen from air} = \frac{80.4 \times 21}{79} = 21.37 \text{ mol}$$

The flue gas contains carbon monoxide in addition to carbon dioxide. Carbon monoxide can consume oxygen for its conversion to carbon dioxide. So the excess oxygen present in the flue gas containing carbon monoxide is termed as apparent excess. To find the actual excess of oxygen, the ThOD of the carbon monoxide (0.5 mole per mole of carbon monoxide) has to be subtracted from the apparent excess of oxygen.

Apparent excess of oxygen = 8.4 mol

Actual excess of oxygen = 8.4 − 0.5 = 7.9 mol

ThOD of the fuel = oxygen supplied − actual excess oxygen
= 21.37 − 7.9 = 13.47 mol

$$\text{Percentage excess of oxygen} = (\text{excess oxygen} \times 100) \div \text{ThOD} = \frac{7.9 \times 100}{13.47} = 58.65\%$$

100 mol of flue gas contains 10.2 mol of carbon dioxide and 1.0 mol of carbon monoxide. 1 mol of carbon gives 1 mol of carbon dioxide on complete combustion or 1 mol of carbon monoxide on partial combustion.

$C + O_2 \rightarrow CO_2$ or $C + 1/2\, O_2 \rightarrow CO$.

Amount of carbon in the fuel = 11.2 mol

1 mol of carbon requires 1 mol of oxygen for its complete combustion to carbon dioxide. So ThOD of carbon is 11.2 mol.

ThOD of the fuel = ThOD of carbon + ThOD of hydrogen

ThOD of hydrogen = ThOD of the fuel − ThOD of carbon
= 13.47 − 11.2 = 2.27 mol

1 mole of hydrogen consumes 0.5 mol of oxygen.

Amount of hydrogen in the fuel = 4.54 mol (twice the amount of oxygen consumed)

Hydrogen is a diatomic molecule and 1 mol of hydrogen contains $2N$ atoms, where N is the Avogadro number.

Number of atoms of hydrogen in the fuel $= 9.08 \times N$ atoms

Carbon is an element and 1 mol carbon contains N atoms.

Number of atoms of carbon in the fuel $= 11.2 \times N$ atoms

Atomic ratio of hydrogen to carbon $= (9.08 \times N) \div (11.2 \times N) = 0.81$

8M A fuel composed of ethane and methane in an unknown proportion is burnt with oxygen enriched air (50mol% oxygen). Orsat analysis of the flue gas reads 25% carbon dioxide, 60% nitrogen and 15% oxygen. Find the composition of the fuel and the mole ratio of enriched air to fuel.

Solution

Basis: 100 moles of dry flue gas. We do not have the entire information on the input side; for example the percentage of excess air supplied is not known. So the problem has to be solved choosing the output as the basis.

Nitrogen in 100 moles of dry flue gas = 60 mol

The fuel is a hydrocarbon mixture. The source of nitrogen in the flue gas is oxygen enriched air supplied for combustion, which contains 50mol% oxygen and 50mol% nitrogen.

Oxygen supplied = 60 mol

Oxygen enriched air supplied = 60 + 60 = 120 mol

ThOD = oxygen supplied – excess oxygen in flue gas = 60 – 15 = 45 mol

Fuel is a hydrocarbon mixture. The products of combustion are carbon dioxide and water vapour.

1 mol of carbon consumes 1 mol of oxygen to give 1 mol of carbon dioxide.

Carbon dioxide in 100 mol of dry flue gas = 25 mol

Oxygen used for carbon dioxide formation = 25 mol

Balance oxygen should have been consumed for water vapour formation.

Oxygen used for water vapour formation = 45 – 25 = 20 mol

1 mol of hydrogen consumes 0.5 mol of oxygen to give 1 mol of water vapour.

Water vapour formed = 40 mol

Let the fuel be composed of X moles of methane and Y moles of ethane.

The combustion reactions are

$$CH_4 + 2O_2 \rightarrow CO_2 + 2H_2O$$

$$C_2H_6 + 3.5O_2 \rightarrow 2CO_2 + 3H_2O$$

Carbon dioxide formed $= X + 2Y = 25$ mol

Water vapour formed $= 2X + 3Y = 40$ mol

$X = 5$ and $Y = 10$

The composition of fuel is given in Table 8.12.

Table 8.12 Composition of fuel

Constituent	Amount (mol)	Mole per cent
Methane	5.0	33.33
Ethane	10.0	66.67
Total fuel	15.0	

Air to fuel ratio $= 120 \div 15 = 8$

8N The analysis of a flue gas obtained from the combustion of a hydrocarbon is carbon dioxide 4.62%, carbon monoxide 3.90%, oxygen 8.92% and nitrogen 82.56%. Compute (i) the number of moles of dry air supplied per mole of dry flue gas, and (ii) the percentage of excess air, if the fuel is burnt completely.

Solution

Basis: 100 mol of dry flue gas. We do not have the entire information on the input side; for example the percentage of excess air supplied is not known. So the problem has to be solved choosing the output as the basis.

Nitrogen in 100 mol of dry flue gas $= 82.56$ mol

The fuel is a hydrocarbon. Source of nitorgen in flue gas is the air supplied for combustion. Air contains 21mol% oxygen and 79mol% nitrogen.

Oxygen accompanying nitrogen $= \dfrac{82.56 \times 21}{79} = 21.95$ mol

Air fed $=$ nitrogen in flue gas + the accompanying oxygen

$= 82.56 + 21.95 = 104.51$ mol

Amount of dry air per mole of dry flue gas $= 104.51 \div 100 = 1.0451$ mol

Flue gas contains carbon monoxide in addition to carbon dioxide. Carbon monoxide can consume oxygen for its conversion to carbon dioxide. So the excess oxygen present in the flue gas containing carbon monoxide is only an apparent excess. To find the actual excess of oxygen, the ThOD of the carbon monoxide (0.5 mol per mole of carbon monoxide) has to be subtracted from the apparent excess oxygen.

Apparent excess of oxygen $= 8.92$ mol

Carbon monoxide in the dry flue gas $= 3.90$ mol

Actual excess of oxygen $= 8.92 - 1.95 = 6.97$ mol

ThOD $=$ oxygen supplied $-$ excess oxygen

$= 21.95 - 6.97 = 14.98$ mol

$$\text{Percentage of excess oxygen} = \frac{\text{excess oxygen} \times 100}{\text{ThOD}} = \frac{6.97 \times 100}{14.98}$$

$= 46.53\%$

80 A furnace using hydrocarbon fuel oil on dry stack gas analysis gave the following result: Carbon dioxide 10.2%, oxygen 8.3% and nitrogen 81.5%. Find (i) the composition of the oil, and (ii) the percentage of excess air used and (iii) the volume of air used per kilogram of fuel oil.

Solution

Basis: 100 kmol of dry flue gas. We do not have the entire information on the input side; for example the percentage excess of air supplied is not known. So the problem has to be solved choosing the output as the basis.

Nitrogen in 100 kmol of dry flue gas $= 81.5$ kmol

The fuel is a hydrocarbon. The source of nitrogen in the flue gas is the air supplied for combustion. Air contains 21mol% oxygen and 79mol% nitrogen.

$$\text{Oxygen accompanying nitrogen} = \frac{81.5 \times 21}{79} = 21.66 \text{ kmol}$$

Air supplied $=$ nitrogen in flue gas $+$ the accompanying oxygen

$= 81.5 + 21.66$

$= 103.16$ kmol

The flue gas does not contain carbon monoxide. So the oxygen present in the flue gas is the actual excess of oxygen.

Excess oxygen $= 8.30$ kmol

ThOD $=$ oxygen supplied $-$ excess oxygen

$= 21.66 - 8.30 = 13.36$ kmol

$$\text{Percentage of excess oxygen} = \frac{\text{excess oxygen} \times 100}{\text{ThOD}} = \frac{8.3 \times 100}{13.36}$$

17 $= 62.12\%$

100 kmol of flue gas contain 10.2 kmol of carbon dioxide. 1 mol of carbon gives 1 mol of carbon dioxide on complete combustion.

$$C + O_2 \rightarrow CO_2$$

Amount of carbon in the fuel = 10.2 kmol

ThOD of carbon = 10.2 kmol

ThOD of the fuel = ThOD of carbon + ThOD of hydrogen

ThOD of hydrogen = ThOD of the fuel – ThOD of carbon

= 13.36 – 10.2 = 3.16 kmol

1 mol of hydrogen consumes 0.5 mole of oxygen. Hence, amount of hydrogen in the fuel is twice the amount of oxygen consumed.

Amount of hydrogen in the fuel = 6.32 kmol

The composition of the fuel is given in Table 8.13.

Table 8.13 Composition of the fuel

Constituent	Amount (kmol)	Molecular weight	Weight (kg)	Weight per cent
C	10.2	12	122.4	90.64
H_2	6.32	2	12.64	9.36
			135.04	

Fuel for 100 kmol of dry flue gas = 135.04 kg

Air supplied for 100 kmol of dry flue gas = 103.16 kmol

Air supplied for 1 kg of fuel $= \dfrac{103.16}{135.04} = 0.764$ kmol = 17.11 m^3 at NTP

8P In a lime manufacturing process, pure calcium carbonate is burnt with coke containing 83% carbon. A gas of composition 26.5% carbon dioxide, 5% oxygen and the balance nitrogen is produced. Compute (i) the ratio of lime produced to coke burnt, (ii) the percentage of excess air, and (iii) the amount of stack gases per tonne of lime produced.

Solution

Basis: 100 kmol of flue gas. We do not have the entire information on the input side; for example the percentage of excess air supplied is not known. So the problem has to be solved choosing the output as the basis.

Coke is burnt in air and the heat of reaction is utilised for decomposing limestone. The reactions are

$$C + O_2 \rightarrow CO_2 + \text{heat}$$

$$CaCO_3 \xrightarrow{\Delta} CaO + CO_2$$

The source of carbon dioxide in the flue gas is the coke burnt in air and the limestone calcined. Carbon dioxide obtained from burning coke can be found by an oxygen balance.

Nitrogen in 100 kmol of flue gas = 68.5 kmol

The source of nitrogen in the flue gas is the air supplied for combustion. Air contains 21mol% oxygen and 79mol% nitrogen.

Oxygen accompanying nitrogen in air $= \dfrac{68.5 \times 21}{79} = 18.2\,\text{kmol}$

Excess of oxygen in 100 kmol of flue gas = 5.0 kmol

Oxygen consumed by burning coke = oxygen supplied − excess oxygen

$= 18.2 - 5.0 = 13.2\,\text{kmol}$

1 mol of carbon consumes 1 mol of oxygen to give 1 mol of carbon dioxide.

Carbon dioxide obtained by burning coke = 13.2 kmol

Carbon required for 13.2 kmol carbon dioxide $= 13.2 \times 12 = 158.4\,\text{kg}$

Carbon content of coke is 83%.

Coke required $= \dfrac{158.4 \times 100}{83} = 190.8\,\text{kg}$

Carbon dioxide from calcium carbonate = carbon dioxide in flue gas − carbon dioxide from coke

$= 26.5 - 13.2 = 13.3\,\text{kmol}$

Limestone on calcination gives 1 kmol carbon dioxide and 1 kmol or 56 kg lime.

Lime produced $= 13.3\,\text{kmol} = 13.3 \times 56 = 744.8\,\text{kg}$

Burning of 190.8 kg of coke releases sufficient heat for the production of 744.8 kg of lime.

Lime to coke ratio $= \dfrac{744.8}{190.8} = 3.9$

5.0 kmol excess oxygen is used for the production of 13.2 kmol carbon dioxide by burning coke.

Percentage of excess oxygen $= \dfrac{5 \times 100}{13.2} = 37.98\%$

Stack gas accompanying the production of 744.8 kg lime is 100 kmol.

$$\text{Stack gas per tonne (1000 kg) of lime} = \frac{100 \times 1000}{744.8} = 134.26 \text{ kmol}$$

8Q A fuel oil with the following composition, carbon 84%, hydrogen 13%, sulphur 1%, oxygen 1%, and water 1% is burnt with excess air. The composition of the flue gas is analysed as: Carbon dioxide 9.9%, carbon monoxide 1.6%, water vapour 10.8%, sulphur dioxide 0.05%, oxygen 3.7% and nitrogen 73.95%. Calculate the percentage of excess air.

Solution

Basis: 100 kg of coal

The combustion reactions are

$$C + O_2 \rightarrow CO_2$$

$$H_2 + 1/2\, O_2 \rightarrow H_2O$$

$$S + O_2 \rightarrow SO_2$$

1 mol of carbon consumes 1 mol of oxygen to give 1 mol of carbon dioxide. 1 mol of hydrogen consumes 0.5 mol of oxygen to give 1 mol of water vapour. 1 mol of sulphur consumes 1 mol of oxygen to give 1 mol of sulphur dioxide.

ThOD is calculated by summing up the oxygen requirement for each combustible in the fuel and subtracting from that the oxygen present in the fuel, as shown in Table 8.14.

Table 8.14 ThOD of the fuel

Constituent	Weight (kg)	Amount (kmol)	ThOD	CO_2 formed (kmol)
C	84	7.0	+7.0	7.0
H_2	13	6.5	+3.250	
S	1	0.031 25	+0.031 25	
O_2	1	0.031 25	−0.031 25	
			10.25	

Flue gas contains carbon monoxide in addition to carbon dioxide. Carbon monoxide can consume oxygen for its conversion to carbon dioxide. So the excess oxygen present in the flue gas containing carbon monoxide is only an apparent excess. To find the actual excess of oxygen, the ThOD of the carbon monoxide (0.5 mol per mole of cabon monoxide) has to be subtracted from the apparent excess of oxygen.

Apparent excess of oxygen for 100 kmol of dry flue gas = 3.70 kmol

Carbon monoxide in 100 kmol of dry flue gas = 1.60 kmol

Actual excess of oxygen for 100 kmol of dry flue gas $= 3.70 - (1.6 \div 2)$

$= 2.90$ kmol

Total carbon oxides in 100 kmol of flue gas $= 9.9 + 1.6 = 11.5$ kmol

If 100 kg of coal is burnt, carbon oxides formed = 7.0 kmol

Flue gas obtained by burning 100 kg of coal $= \dfrac{100 \times 7}{11.5} = 60.87$ kmol

Actual excess of oxygen in 60.87 kmol of flue gas = 1.77 kmol

Percentage of excess oxygen $= \dfrac{\text{excess oxygen} \times 100}{\text{ThOD}}$

$= \dfrac{1.77 \times 100}{10.25} = 17.27\%$

8R A coal containing 68% carbon, 4% hydrogen, 1% sulphur, 10% nitrogen, 8% oxygen and the rest ash is burnt in a furnace. Orsat analysis of the product of combustion reads carbon dioxide 15%, oxygen 4.7% and the rest nitrogen. Calculate (i) the volume of air used at NTP for the complete combustion of 100 kg of coal, and (ii) the percentage of excess air used.

Solution

Basis: 100 kg of coal

The combustion reactions are

$$C + O_2 \rightarrow CO_2$$

$$H_2 + 1/2\, O_2 \rightarrow H_2O$$

$$S + O_2 \rightarrow SO_2$$

1 mol of carbon consumes 1 mol of oxygen to give 1 mol of carbon dioxide. 1 mol of hydrogen consumes 0.5 mol of oxygen to give 1 mol of water vapour. 1 mol of sulphur consumes 1 mol of oxygen to give 1 mol of sulphur dioxide.

ThOD is calculated by summing up the oxygen requirement for each combustible in the fuel and subtracting from that the oxygen present in the fuel, as given in Table 8.15.

Basis: 100 kmol of dry flue gas

Orsat reads carbon dioxide and sulphur dioxide as carbon dioxide and the individual amounts of these two gases is to be calculated based on their elemental ratio.

Table 8.15 ThOD of the fuel

Constituent	Weight (kg)	Amount (kmol)	Oxygen demand (kmol)
C	68	5.667	+ 5.667
H_2	4	2	+ 1.000
S	1	0.031 25	+ 0.031 25
O_2	8	0.25	− 0.25
N_2	10	0.357	
		ThOD	6.448

Oxides of carbon and sulphur present in 100 kmol of dry flue gas = 15 kmol

In 100 kg of coal, 5.667 kmol of carbon and 0.031 25 kmol of sulphur are present. Out of 5.698 kmol of carbon and sulphur, 5.667 kmol is carbon. Oxide formed from the elements also should be in the same ratio.

Carbon dioxide in 100 kmol dry flue gas $= \dfrac{15 \times 5.667}{5.698} = 14.92$ kmol

1 kmol of carbon is required for 1 kmol of carbon dioxide.

Carbon required for 100 kmol of dry flue gas = 14.92×12 kg

Coal contains 68% carbon by weight.

Coal required for 100 kmol of dry flue gas $= \dfrac{14.92 \times 12}{0.68}$

$= 263.27$ kg

The source of nitrogen in flue gas is air and coal.

Amount of nitrogen in the flue gas = 80.3 kmol

Amount of nitrogen in the coal $= \dfrac{263.27 \times 0.1}{28} = 0.94$ kmol

Nitrogen from air = nitrogen in flue gas − nitrogen in coal

$= 80.30 - 0.94 = 79.36$ kmol

Oxygen accompanying nitrogen in air $= \dfrac{79.36 \times 21}{79} = 21.10$ mol

Air supplied = nitrogen in air + accompanying oxygen

$= 79.36 + 21.10 = 100.46$ kmol

Air supplied for 263.27 kg coal = 100.46 kmol

$$\text{Air supplied for 100 kg coal} = \frac{100.46 \times 100}{263.27} = 38.16\,\text{kmol}$$

$$= 854.75\,\text{m}^3 \text{ at NTP}$$

Air contains 21mol% of oxygen.

Oxygen supplied for 100 kg coal = $38.16 \times 0.21 = 8.015$ kmol

ThOD = 6.448 kmol

$$\text{Percentage of excess oxygen} = \frac{\text{excess oxygen} \times 100}{\text{ThOD}} = \frac{1.567 \times 100}{6.448} = 24.30\%$$

8S Natural gas containing 85% methane, 3% propane, 3% benzene, 5% hydrogen and the rest water vapour is burnt with 20% excess air. Air is 1.5% moist. Calculate (i) the volume of air required for the combustion of one cubicmetre of natural gas, and (ii) the composition of dry flue gas produced, assuming that combustion is 100% complete.

Solution

Basis: 100 kmol of natural gas

The combustion reactions are:

$$CH_4 + 2O_2 \rightarrow CO_2 + 2H_2O$$

$$C_2H_6 + 7/2\,O_2 \rightarrow 2CO_2 + 3H_2O$$

$$C_6H_6 + 15/2\,O_2 \rightarrow 6CO_2 + 3\,H_2O$$

$$H_2 + 1/2\,O_2 \rightarrow H_2O$$

ThOD and the amount of carbon dioxide and water vapour formed for each of the constituents of natural gas are calculated making use of the above equations and tabulated as given in Table 8.16.

Table 8.16 ThOD of the fuel

Constituent	Amount (kmol)	O_2 required per kilomole (kmol)	ThOD	CO_2 formed (kmol)	H_2O formed (kmol)
Methane	85	2	170	85	170
Propane	3	3.5	10.5	6	9
Benzene	3	7.5	22.5	18	9
Hydrogen	5	0.5	2.5	–	5
			205.5	109	193

20% excess of oxygen is supplied.

$$\text{Oxygen supplied} = \frac{205.5 \times 120}{100} = 246.6\,\text{kmol}$$

The source of oxygen is air.

$$\text{Nitrogen accompanying oxygen in air} = \frac{246.6 \times 79}{21} = 927.2\,\text{kmol}$$

Air supplied = oxygen supplied + accompanying nitrogen

= 246.6 + 927.2 = 1 173.8 kmol

Air supplied is moist and its moisture content is 1.5%.

$$\text{Moist air supplied} = \frac{1\,173.8 \times 101.5}{100} = 1\,191.4\,\text{kmol}$$

For gases, number of moles and volume are analogous.

Volume of air supplied for 1 m^3 natural gas = 1 191.4 ÷ 100 = 11.914 m^3 at same temperature and pressure.

Excess of oxygen = oxygen supplied − ThOD = 246.6 − 205.5 = 41.1 kmol

The composition of the dry flue gas is given in Table 8.17.

Table 8.17 Composition of dry flue gas

Constituent	Amount (kmol)	Mol%
CO_2	109.0	10.12%
O_2	41.1	3.82%
N_2	927.2	86.06%
	1077.3	

8T Natural gas containing 78.8% methane, 14% ethane, 0.4% carbon dioxide and 6.8% nitrogen by volume is burnt with 40% excess of air. The hydrocarbons are burnt to give 74% carbon dioxide and 26% carbon monoxide. Calculate (i) the Orsat analysis of the flue gas, and (ii) the volume of the stack gas produced for 1 m^3 of natural gas.

Solution

Basis: 100 kmol natural gas

The combustion reactions are:

$$CH_4 + 2O_2 \rightarrow CO_2 + 2H_2O$$

$$CH_4 + 3/2\,O_2 \rightarrow CO + 2H_2O$$

$$C_2H_6 + 7/2\,O_2 \rightarrow 2CO_2 + 3H_2O$$
$$C_2H_6 + 5/2\,O_2 \rightarrow 2CO + 3H_2O$$

ThOD is to be determined based on complete combustion.

ThOD = (78.8 × 2) + (14 × 3.5) = 206.6 kmol

40% excess of oxygen is supplied. The source of oxygen is air. Air contains 21mol% oxygen and 79mol% nitrogen.

Actual oxygen supplied = 206.6 × 1.4 = 289.24 kmol

$$\text{Nitrogen accompanying oxygen in air} = \frac{289.24 \times 79}{21} = 1\,088.09\ \text{kmol}$$

The hydrocarbons are burnt to yield 74% carbon dioxide and 26% carbon monoxide. The amount of products of combustion can be calculated making use of the combustion equations given above.

Combustion of methane

Carbon dioxide formed = 78.8 × 0.74 × 1 = 58.312 kmol

Carbon monoxide formed = 78.8 × 0.26 × 1 = 20.488 kmol

Water vapour formed = 78 × 2 = 157.6 kmol

Combustion of ethane

Carbon dioxide formed = 14 × 0.74 × 2 = 20.72 kmol

Carbon monoxide formed = 14 × 0.262 × 1 = 7.28 mol

Water vapour formed = 14 × 3 = 42 kmol

Total carbon dioxide formed = 58.312 + 20.72 = 79.032 kmol

Total carbon monoxide formed = 20.488 + 7.28 = 27.768 kmol

Total water vapour formed = 4157.6 + 42 = 199.6 kmol

Each mole of carbon dioxide requires 1 mol of oxygen and each mole of carbon monoxide as well as water vapour require 0.5 mol of oxygen. Using this the oxygen consumed can be calculated.

Oxygen consumed = 79.032 + (27.768 ÷ 2) + (199.6 ÷ 2) = 192.72 kmol

Excess oxygen = oxygen supplied − oxygen consumed
= 289.24 − 192.72 = 96.52 kmol

Total carbon dioxide in flue gas = carbon dioxide formed + carbon dioxide in fuel
= 79.032 + 0.4 = 79.432 kmol

Total nitrogen in flue gas = nitrogen from air + nitrogen in fuel
= 1 088.09 + 6.8 = 1 094.89 kmol

The Orsat analysis data is given in Table 8.18.

Table 8.18 Orsat analysis

Constituent	Amount (kmol)	Volume per cent
CO_2	79.432	6.12
CO	27.768	2.14
O_2	96.52	7.43
N_2	1 094.89	84.31
	1 298.61	

Stack gas produced for 100 kmoles of fuel $=$ dry flue gas $+$ water vapour

$= 1\,298.61 + 199.60 = 1\,498.21$ kmol

Amount of flue gas formed per kilomole of natural gas $= 1\,498.21 \div 100 = 14.982\,1$ kmol

For gases, composition by mole and volume are analogous.

Volume of flue gas formed for 1 m^3 natural gas $= 14.982\,1$ m^3

Problems for Practice

8.1 Crude oil is found to contain 87% carbon, 12.5% hydrogen and 0.5% sulphur by weight. If its gross calorific value is 10 800 kcal kg^{-1} at 298 K, calculate the net calorific value at 298 K.

8.2 A coal containing 87% carbon, 7% hydrogen and the rest ash is burnt with 40% excess air. Calculate the weight of air used for 1 kg of coal burnt.

8.3 The gross calorific value of pure ethane is 372.82 kcal mol^{-1} at 298 K. Find its net calorific value at the same temperature.

8.4 30 kg of coal, analysing 80% carbon and 20% hydrogen, are burnt with 600 kg of air to yield a gas having an Orsat analysis in which the ratio of carbon dioxide to carbon monoxide is 3:2. What is the percentage excess of air?

8.5 10 kmol of sulphur are burnt with 100 kmol of air. 90% of sulphur is burnt to sulphur dioxide and the rest to sulphur trioxide. Compute the flue gas analysis on sulphur trioxide-free basis.

8.6 In the manufacture of sulphur trioxide, feed to the reactor consists of 50 kmol sulphur dioxide and 150 kmol air. Compute the percentage of excess air used.

8.7 A combustion chamber is fed with 50 kmol of butane and 2 000 kmol of air per hour. Calculate the percentage excess of air used.

8.8 A natural gas containing 95% methane and 5% ethane is burnt with 20% excess air. Compute the weight of air required per kilomole of fuel.

8.9 Butane is burnt with 50% excess air: 70% butane is burnt to carbon dioxide, 20% is burnt to carbon monoxide and the rest remains unburnt. Determine the composition of the flue gases formed.

8.10 A furnace uses natural gas, which consists entirely of hydrocarbons. The dry flue gas analysis shows carbon dioxide 9.5%, carbon monoxide 1.9%, oxygen 1.4% and the rest nitrogen. Calculate (i) the atomic ratio of carbon:hydrogen, (ii) the excess air used, (iii) the composition of fuel, and (iv) the air to fuel ratio.

8.11 A petroleum refinery burns a gas mixture containing 7% pentane, 10% butane, 16% propane, 9% ethane, 55% methane and 3% nitrogen at the rate of 200 $m^3 h^{-1}$ at 30°C and 4.5 bar pressure. The air rate is adjusted so that 15% excess air is used. Under these conditions the molar ratio of carbon dioxide to carbon monoxide in the flue gas is 20:1. Calculate (i) the volume of air introduced every hour at 303 K and 1 atm pressure, and (ii) the composition of flue gas on dry basis.

8.12 The Orsat analysis of flue gas from an oil furnace is carbon dioxide 8%, carbon monoxide 3%, oxygen 4% and nitrogen 85%. Analysis indicates that oil contains 78% by weight of carbon and the rest combustible hydrogen and moisture. Dry air enters at 303 K and 1.013 bar. Calculate (i) the percentage of excess air used, (ii) the carbon to hydrogen ratio in the fuel oil, (iii) the volume of air used per 100 kg of oil fired, and (iv) the weight of moisture in the flue gas per kilogram of oil fired.

8.13 A fuel oil consisting of carbon 88.2% and hydrogen 11.8% is burnt with 20% excess air. 95% of carbon is burnt to carbon dioxide and the rest to carbon monoxide. All the hydrogen is burnt to water. Determine Orsat analysis of the flue gases formed.

8.14 A gas containing 80% ethane and 20% oxygen is burnt with 200% excess air. Only 80% of ethane is burnt to carbon dioxide, 10% to carbon monoxide and the rest remains unburnt. Compute the stack gas composition on dry basis.

8.15 A producer gas made from coke has the following composition by volume: Carbon monoxide 28%, carbon dioxide 3.5%, oxygen 0.5% and nitrogen 68%. The gas is burnt with such a quantity of air that the oxygen from the air is 20% in excess of the net oxygen required for complete combustion. If the combustion is 98% complete, calculate the composition of the gaseous products formed per 100 kg of gas burnt.

8.16 A furnace uses coke containing 80% carbon, 5% hydrogen and the remaining ash. The furnace operates with 50% more air than that required for complete combustion of all the carbon charged. The solid residue contains 2% carbon. Of the carbon burnt, 5% goes to carbon monoxide. Calculate (i) the composition of the flue gas, (ii) the ash produced, and (iii) the weight of carbon lost for 1 kg of coke burnt.

8.17 A producer gas analyses as follows: Hydrogen 15%, methane 3%, ethylene 1%, carbon dioxide 10%, carbon monoxide 15% and the rest nitrogen. Compute (i) the theoretical air required for combustion, and (ii) the composition of the flue gas if the gas is burnt with 50% excess oxygen.

8.18 Producer gas containing 14% hydrogen, 24% carbon monoxide, 7% carbon dioxide, 5% methane and the rest nitrogen is burned with 25% excess air. Calculate (i) the flue gas analysis, and (ii) the volume of flue gas leaving the burner at 400°C and 730 mm Hg pressure, for 1 kmol of producer gas burnt.

8.19 A gas having 82% methane, 16.5 % ethane and the rest butane is burned with 60% excess air. Assuming 70% combustion, find the exit gas analysis.

8.20 A hydrocarbon fuel with the general formula C_xH_y is burnt in excess air, with air being fed to the furnace at 800 mm Hg and 30°C. Analysis of exit gases shows carbon dioxide 12.28%, carbon monoxide 3.07%, oxygen 2.34%, water vapour 19.18% and nitrogen 63.13%. Calculate (i) the percentage of excess oxygen fed, and (ii) the volume of air fed for 1 kmol of fuel gas.

8.21 A pure hydrocarbon gas is burnt in a furnace giving a flue gas containing 10.8% carbon dioxide, 3.8% oxygen and the rest nitrogen. Compute the atomic ratio of carbon to hydrogen and percentage of excess air used.

8.22 Natural gas containing 92% methane, 6.2% ethane and the balance propane is burnt with 100% excess air. Assuming complete combustion, calculate the Orsat analysis of the product.

8.23 Methane burns to carbon dioxide and water vapour. Even when 20% excess air is used, the combustion is only 70% complete. Find the dry exit gas analysis.

8.24 Coke containing 89% carbon and the rest ash is burnt in a furnace. The grate efficiency is such that 90% of the carbon present in the coke charged is burnt. 30% excess air is used. 97% of carbon burnt is oxidised to carbon dioxide, while the remainder goes to carbon monoxide. Determine the composition of the gases leaving the furnace. If coke is fed at the rate of 100 kg h^{-1}, estimate the flow rate of the flue gases at 298 K and 1 atm pressure.

8.25 A producer gas contains 9.2% carbon dioxide, 0.4% ethylene, 20.9% carbon monoxide, 15.6% hydrogen, 1.9% methane and 52% nitrogen. When it is burnt, the products of combustion are found to contain carbon dioxide 10.8%, carbon monoxide 0.4%, oxygen 9.2% and nitrogen 79.6%. Calculate (i) the volume of air required for 1 m^3 of producer gas, (ii) the excess of air used and (iii) the percentage of nitrogen that has come from the fuel.

8.26 A natural gas containing 95% methane and 5% nitrogen is burnt with excess air. Orsat analysis of the flue gases reads 9.1% carbon dioxide, 0.2% carbon monoxide, 4.6% oxygen and 86.1% nitrogen. Compute the percentage of excess air.

8.27 A producer gas made from coal has the following composition by volume: Carbon monoxide 16%, carbon dioxide 4%, hydrogen 50%, methane 15%, ethane 3%, benzene 2%, and the balance nitrogen. It is burnt with 20% excess air. Assume complete combustion. Calculate (i) the amount of air required per 100 kmol of gas, (ii) the amount of flue gases produced per 100 kmol of gas, and (iii) the composition of the dry products of combustion.

Answers to Problems

CHAPTER 1

1.1 (i) $3.664 \times 10^{-3}\ m^2 s^{-1}$

(ii) 344.61 kN m^{-2}

(iii) 105.5 kJ

(iv) 1.34 HP

(v) 16 018 kg m^{-3}

(vi) 310.78 K

1.2 0.271 2 kg $m^{-2}\ s^{-1}$

1.3 $h_y = 1.135 \times 10^{-5} \dfrac{G^{0.8}}{D_e^{0.2}}$

1.4 $kg_a = 0.012\, G_y^{0.7} G_x^{0.25}$

1.5 1 703.49 to 5 678.3 W $m^{-2}\ K^{-1}$

CHAPTER 2

2.1 179.03 m^3

2.2 46.93%

2.3 97.14%

2.4 33 680 mol

2.5 200 kPa

2.6 0.192 4 kg

2.7 8.75 atm

2.8 17.47 m^3

2.9 615.85 K

2.10 30.40

2.11 52

2.12 40%

2.13 8.97 N, 8.97 M, molality = 10.71

2.14 16.49wt%

2.15 6.93 N, 6.93 M, molality = 10.83

2.16 Nitric acid bottle contains 4.788×10^{23} atoms more

2.17 365 kg

2.18 209.7 kg m^{-3}

2.19 188.57 g

2.20 (i) 0.071, 3.96 N, 3.95 M

(ii) Molality 1.427; 92.3 vol%

2.21 (i)

Constituent	Wt%
Zinc bromide	64.9
Water	35.1

(ii)

Constituent	Vol%
Zinc bromide	12.88
Water	87.12

(iii)

Constituent	At%
Zinc	4.29
Bromine	8.58
Hydrogen	58.08
Oxygen	29.05

(iv) 8.212 m

(v) 5.773 M

2.22

Constituent	Mol%
Pentane	46.45
Hexane	24.31
Heptane	29.24

Average molecular weight 83.54

2.23

Constituent	Mol%
Benzene	41.67
Toluene	35.34
Xylene	22.99

Average molecular weight 89.36

2.24 5.015

2.25 86.76%

2.26 49.36 m^3

2.27 54.91 m^3

2.28 (i)

Constituent	Mol%
Nitrogen	67.58
Oxygen	21.39
Ammonia	5.68
Water vapour	5.35

(ii) 2.418 m^3

2.29 (i) 0.1875

(ii) 2.475 atm

(iii) 2.125×10^{23} atoms

2.30 (i)

Constituent	Wt%
Methane	60.94
Ethane	17.58
Propane	21.48

(ii) 20.48

(iii) 0.5746 kg m^{-3}

(iv) 0.7062

2.31 (i)

Constituent	Mol%
Chlorine	65.62
Bromine	21.15
Oxygen	13.23

(ii) 3.64 kg m^{-3}

(iii) 2.92

2.32

Constituent	(i) Partial pressure (bar)	(ii) Partial volume (m^3)
Hydrogen sulphide	0.12	0.187
Butane	0.16	0.250
Propane	0.12	0.187
Ethane	0.12	0.187
Methane	3.28	5.111
Helium	0.08	0.125
Nitrogen	0.12	0.187

(iii) Density 3.144 kg m^{-3}

2.33

Constituent	Vapour phase (mol%)	Liquid phase (mol%)
Benzene	79.29	55.45
Toluene	16.98	28.20
o-Xylene	3.73	16.35

Total pressure = 1.231 bar

2.34 Nitrogen: 20.78 m^3 min^{-1};
Hydrogen: 24.94 m^3 min^{-1}

CHAPTER 3

3.1 2247.45 m^3

3.2 21.17 m^3, 0.27 kg

3.3 (i) 107.18 L

(ii) 23.94 g

3.4 A: 33.33%, B: 66.67%

3.5 320 kg of 25% sodium chloride solution, 280 kg of 30% potassium chloride solution and 400 kg of mixed solution of 5% sodium chloride and 4% potassium chloride

3.6

90% Nitric acid	2833 kg
95% Sulphuric acid	5366 kg
Waste acid	1801 kg

3.7

98% Sulphuric acid	336.91 kg
90% Nitric acid	313.94 kg
Waste Acid	349.15 kg

3.8

99% Glycerol	479.8 kg
Iodine	9.752 kg
Water	10.45 kg

3.9 410.57 kmol h^{-1}

3.10 54.29% oxygen and 45.71% helium

3.11 (i) 6863.64 kg
(ii) 1023.4 kg

3.12 432 mol min^{-1}

3.13 0.56 Tonne

CHAPTER 4

4.1

Constituent	Vol%
Sulphur dioxide	15.22
Sulphur trioxide	0.80
Oxygen	4.67
Nitrogen	79.31

Volume = 4665.6 m^3

4.2 64.2 h

4.3 (i) 1528.87 kg
(ii) 1510.66 kg

Composition	Wt%
Nitric acid	1.6
Sulphuric acid	72.62
Water	25.78

4.4 (i) 1.4283 kg

(ii)

Constituent	Mol%
Hydrogen chloride	35.58
Oxygen	13.38
Nitrogen	51.04

(iii)

Constituent	Mol%
Hydrogen chloride	15.09
Chlorine	11.25
Water vapour	11.25
Oxygen	8.5
Nitrogen	53.91

4.5 Antimony sulphide is the limiting reagent
Iron is the excess reagent

Percentage excess	= 46.76
Degree of completion	= 69.72%
Conversion of iron	= 47.51%

4.6 (i) 0.0716
(ii) 16.3%
(iii) 2303.44 m^3 h^{-1}

4.7 (i) 26.71%
(ii) 95.97%
(iii) 0.548 kg

4.8

Composition	Burner gas (mol%)	Converter gas (mol%)
Sulphur dioxide	8.31	4.30
Sulphur trioxide	3.56	7.82
Oxygen	7.22	5.79
Nitrogen	80.91	82.09

4.9 (i) 1 100.11 kg
(ii) Hydrogen chloride gas removed: 280.65 kg
Water removed: 55.01 kg

(iii)

Constituent	Wt%
Sodium chloride	28.34
Sodium bisulphate	58.17
Hydrochloric acid	1.77
Water	9.35
Sulphuric acid	2.39

4.10 63 kg

4.11 (i)

Constituent	Wt%
Calcium phosphate	52.36
Silica	33.45
Coke	14.19

(ii) 6.6 kg

4.12 (i) 310.6 m^3
(ii) 69.18
(iii) 881.5 m^3

4.13 Benzene: 746.07 kg;
Nitric acid: 994.27 kg

4.14 89.29%

4.15 (i) 30%
(ii) 4.9%
(iii) 1:7

4.16 Benzene: 2 063.5 kg h^{-1};
Hydrogen: 208.04 kg h^{-1}

4.17 *o*-Xylene: 1 582.36 kg;
Air: 7 466.34 kg

4.18

Constituent	Mol%
Toluene	8.27
Benzaldehyde	1.19
Oxygen	17.38
Carbon dioxide	0.33
Water vapour	1.38
Nitrogen	71.45

4.19 (i) 22879.82 $m^3\ h^{-1}$
(ii) 61861.24 $m^3\ h^{-1}$

4.20 (i) 4472.9 m^3
(ii) 13418.7 m^3

4.21 0.571

4.22 3.4 kmol per 100 kmol fresh feed

4.23 (i) 292.46 kmol h^{-1}
(ii) 198.1 kmol h^{-1}
(iii)

Constituent	Mol%
Aniline	13.31
Nitrobenzene	0.56
Cyclohexylamine	0.14
Hydrogen	85.35
Inerts	0.64

4.24 Entrainer required: 2103 kg;
Water evaporated: 650 kg

4.25 0.444

4.26 Volume of entering air:
1061.66 m^3;
Bypass: 20.41% by mole

CHAPTER 5

5.1 (i) I effect: 3000 kg h^{-1}; II effect: 1000 kg h^{-1}
(ii) 96 tonnes per day
(iii) 8000 kg

5.2 9.605 kg h^{-1}; 187.124 m^3

5.3 4.176 wt%

5.4 3.898 kg h^{-1} ; 91.42 $m^3\ h^{-1}$

5.5 Distillate: 370.98 kg; Residue: 629.02 kg

5.6 Distillate: 167.65 kg; Residue: 832.35 kg

Composition of residue

Constituent	Wt%
Water	99.01
Methanol	0.99

5.7

Constituent	Distillate (wt%)	Residue (wt%)
Propane	99.43	0.2
Butane	0.57	99.8

5.8 416.27 kg

5.9 Solution: 552.44 kg containing 13.91% oil and 89.09% solvent

5.10 91.67%

5.11 Rs. 360.50

5.12 1834.19 kg

5.13

Constituent	Weight
Feed	3138.02 kg
Mother liquor	135.94 kg
Evaporation	502.08 kg

5.14

Constituent	Weight
Crystals	4830.8 kg
Mother liquor	4454.5 kg
Evaporation	488.7 kg

5.15

Constituent	Weight
Crystals	2832.95 kg
Mother liquor	10835.23 kg
Evaporation	422.73 kg

5.16

Constituent	Weight
Crystals	6675.1 kg
Mother liquor	2235.2 kg
Evaporation	670.7 kg

5.17 142.62 kg

5.18 0.899; 16.16 kg

5.19 (i) 0.019 mol water per mol DA

(ii) 0.000 5 mol water per mol DA

(iii) 0.388 kg

(iv) 11.52 m^3

5.20 47.064 m^3; 68.31 m^3

5.21 15.732 kg

5.22 33.45 m^3

5.23 978.76 $m^3\ h^{-1}$

5.24 2 126.2 mm Hg

5.25 (i) 0.021 06

(ii) 0.01307

(iii) 51.51

(iv) 50.49

(v) 1.025 kJ per kg dry air per K

(vi) 0.884 m^3 per kg dry air

5.26 493.83 kg

5.27 180.24 mm Hg; 0.897 kg of benzene per kg of nitrogen

CHAPTER 6

6.1 18.1%; 6.932 h

6.2 8.29 h

6.3 4.506 h

6.4 0.3825 kg

6.5 32.19 min

CHAPTER 7

7.1 191.62 kJ

7.2 (i) 35445 kcal

(ii) 32922 kcal

(iii) 35451 kcal

7.3 (i) 6992.95 kcal

(ii) 6993 kcal

7.4 20385.92 kJ $kmol^{-1}$

7.5 1276.25 kJ kg^{-1}

7.6 −903.6 kJ mol^{-1}

7.7 −195 kJ mol^{-1}

7.8 (i) −72 kJ mol^{-1}.

(ii) 1893 kJ mol^{-1}.

7.9 2012.91 kcal

7.10 -41.4 kJ mol^{-1}

7.11 41294.1 kcal

7.12 47885400 J

7.13 1253.7 K

7.14 1242.22 K

7.15 -6.57 kJ kg^{-1}

7.16 1133.32 kcal $kmol^{-1}$ gas

7.17 1137.6 K

CHAPTER 8

8.1 10143.9 kcal kg^{-1}

8.2 17.53 kg

8.3 341.34 kcal kg^{-1}

8.4 21.43%

8.5

Constituent	Mol%
Sulphur dioxide	9.14
Oxygen	10.66
Nitrogen	80.2

8.6 26%

8.7 29.23%

8.8 346.86 kg

8.9

Constituent	Mol%
Butane	0.2
Carbon dioxide	5.7
Carbon monoxide	1.63
Water vapour	9.16
Oxygen	8.74
Nitrogen	74.57

8.10 (i) 1:4

(ii) 1.98%

(iii) Methane

(iv) 9.68%

8.11 (i) 16663.95 m^3 per hour

(ii)

Constituent	Mol%
Carbon dioxide	10.75
Carbon monoxide	0.54
Oxygen	3.23
Nitrogen	85.48

8.12 (i) 12.44%

(ii) Carbon:Hydrogen = 3.6:1 by weight

(iii) 1424.11 m^3 at NTP

(iv) 1.937 kg

8.13

Constituent	Mol%
Carbon dioxide	12.48
Carbon monoxide	0.66
Oxygen	3.95
Nitrogen	82.91

8.14

Constituent	Mol%
Ethane	0.22
Carbon dioxide	3.51
Carbon monoxide	0.44
Oxygen	15.27
Nitrogen	80.56

8.15

Constituent	Mol%
Carbon dioxide	18.93
Carbon monoxide	0.34
Oxygen	1.84
Nitrogen	78.89

8.16 (i)

Constituent	Mol%
Carbon dioxide	10.88
Carbon monoxide	0.57
Water vapour	4.31
Oxygen	7.18
Nitrogen	77.06

(ii) 15.306 kg

(iii) 0.306 kg

8.17 (i) 1.1429 mol per mole of fuel

(ii)

Constituent	Mol%
Carbon dioxide	11.79
Water vapour	8.26
Oxygen	4.72
Nitrogen	75.23

8.18 (i)

Constituent	Mol%
Carbon dioxide	17.65
Water vapour	11.74
Oxygen	2.33
Nitrogen	68.28

(ii) 117.31 m^3 per kilomole of fuel

8.19

Constituent	Mol%
Butane	0.02
Ethane	0.26
Methane	1.32
Carbon dioxide	4.53
Water vapour	8.27
Oxygen	11.14
Nitrogen	74.46

8.20 (i) 5%

(ii) 17.93 m^3 at NTP per kilomole of fuel

8.21 Atomic ratio of Carbon:
Hydrogen = 1 : 3 ; 20.1% excess.

8.22

Constituent	Mol%
Carbon dioxide	5.66
Oxygen	11.07
Nitrogen	83.27

8.23

Constituent	Mol%
Carbon dioxide	6.35
Methane	2.72
Oxygen	9.07
Nitrogen	81.86

8.24

Constituent	Mol%
Carbon dioxide	14.07
Carbon monoxide	0.43
Oxygen	6.67
Nitrogen	78.83

1125.13 $m^3 h^{-1}$

8.25 (i) 2.29 m^3

(ii) 106.45%

(iii) 77.7%

8.26 24.64%

8.27 (i) 505.5 kmol

(ii) 575 kmol

(iii)

Constituent	Mol%
Carbon dioxide	11.04
Oxygen	3.69
Nitrogen	85.27